Fundamental Building Technology

Second Edition

Andrew J. Charlett and
Craig Maybery-Thomas

Routledge
Taylor & Francis Group

LONDON AND NEW YORK

First edition published 2007

This edition published 2013
by Routledge
2 Park Square, Milton Park, Abingdon, Oxon, OX14 4RN

Simultaneously published in the USA and Canada
by Routledge
711 Third Avenue, New York, NY 10017

Routledge is an imprint of the Taylor & Francis Group, an informa business

British Library Cataloguing in Publication Data
A catalogue record for this book is available from the British Library

Library of Congress Cataloging-in-Publication Data

Charlett, Andrew J.
Fundamental building technology / Andrew J. Charlett and Craig Maybery-Thomas. — 2nd ed.
   p. cm.
Includes bibliographical references and index.
1. Building.   I. Maybery-Thomas, Craig.   II. Title.
TH146.C514 2013
690—dc23
2012017856

ISBN13: 978-0-415-69258-8 (hbk)
ISBN13: 978-0-415-69259-5 (pbk)
ISBN13: 978-0-203-15517-2 (ebk)

Typeset in Helvetica
by RefineCatch Limited, Bungay, Suffolk

MIX
Paper from responsible sources
FSC® C004839

Printed and bound in Great Britain by the MPG Books Group

Contents

Illustrations

Fundamental Building Technology

Fundamental Building Technology introduces the technology, methods and processes fundamental to construction by focussing on what is involved in building a typical low-rise house. Written with the novice in mind, this textbook is the ideal starting point for any construction student, as it fully supports the reader all the way to understanding the functional requirements of each element of the building, and how to take these into account through the building process itself.

This second edition is expanded to cover even more relevant topics, and is supported by more resources for use by the student and lecturer. Now included are:

- An introduction to the planning process and the building regulations
- How to incorporate a sustainable approach, in the selection of materials and elsewhere
- A companion website with lecturer's answers manual and illustrated lecture notes
- Over 150 labelled diagrams throughout the book, and multiple self-study questions in every chapter
- A students' section of the companion website with multiple choice quizzes and over 200 full-colour photos linked to chapters of the book.

Concise, focussed and the most student-friendly guide to this topic available, *Fundamental Building Technology* is the perfect textbook for those taking construction technology modules at undergraduate or HNC/HND level.

Andrew J. Charlett was previously Principal Lecturer in the School of Architecture, Design and the Built Environment at Nottingham Trent University, UK. He has almost thirty years' experience of teaching building technology to a range of further and higher education students.

Craig Maybery-Thomas is currently a Land and Measured Building Surveyor working within the Property and Regeneration Section of Neath Port Talbot County Borough Council. He has previously lectured in construction technology and surveying at Swansea Metropolitan University, UK.

Tables

Foreword

When considering the fundamental aspects of building technology, it is important to consider the functional requirements of each element, material or component that make up a building. There is no one single method of constructing a building correctly. Instead there are various techniques that may be considered equally as correct. What matters is whether these techniques satisfy the basic fundamental requirements of construction. This book approaches the study of building technology from the standpoint of developing an understanding of these functional requirements of each part of a building and explaining how each of these requirements may be satisfied. In order to understand these fundamental aspects as easily as possible, the house has been chosen as the basic building model.

The book progresses through the various stages of house construction in a logical sequence and also considers the main materials used. Throughout the book a number of details have been provided to illustrate the various construction techniques being described. It should be appreciated that each of these details is only one example of how the functional requirements of the particular part of the building being considered may be satisfied.

As particular concepts are introduced, self-assessment questions are posed. These enable the reader to test their own understanding of each concept being considered. Having formulated an answer to each of these questions, the reader may consult the website associated with this book for a suggested answer. Again, these answers are not necessarily a definitive solution to each question posed, but it is hoped that they provide a useful understanding of the concepts being considered.

The first edition of this book also introduced a consideration of other important concepts in building technology, such as life cycle costing, dimensional and modular co-ordination and thermal performance. This second edition continues this approach by also focussing on the increasingly important aspect of sustainability. Throughout the book, reference is made to the requirements of the Building Regulations, 2010 edition, and references are also made to the main considerations of many of the Approved Documents, particularly those in Approved Document L1A, Conservation of fuel and power (new dwellings). New materials and techniques are also considered. The website associated with this book has also been expanded to include lecture notes, multiple choice test questions and a gallery of colour photographs of various aspects of house construction.

Acknowledgements

We would like to thank the following companies for kindly granting us permission to use the following illustrations:

Caterpillar Inc. for Figures 4.4–4.11 inclusive
Krings International for Figure 4.17

Permission to reproduce extracts from British Standards for Tables 3.1, 3.2, 3.3, 7.1 and 10.3 is granted by the British Standards Institution.

Permission to use extracts from BRE Digest DG 429 for Table 10.2 is granted by the Building Research Establishment.

CHAPTER

An introduction to building technology

A building consists of an assembly of materials and components, joined together in such a way as to allow the building to fulfil its primary purpose, that of providing shelter to its occupants. This primary function may be augmented by other supplementary functions; for instance, the building may be used for certain activities, a factory will be used for the manufacture of products, a warehouse will be used for the storage of goods, a leisure centre will be used for the pursuit of leisure activities, whilst a concert hall will be used to house entertainment activities. Nevertheless, they will all need to satisfy the primary objective of providing shelter to the occupants.

The provision of shelter rests on two basic necessities:

- that the building will act as an enclosure for the activities housed within;
- that the building will protect the occupants, equipment or goods housed within from the vagaries of the external climate (rain, wind, sun, snow and frost).

Building elements

If the building is to act as an enclosure, it must have external walls and be covered by a roof. The roof will normally rest on the walls and be supported by them. The walls, in turn, will need a firm base or foundation to be built upon, which will transfer their weight and that of the roof to the ground beneath.

In order to make the building useable, the internal space enclosed by the external walls and roof may need to be sub-divided into rooms by the introduction of horizontal dividers between storeys – the floors, and vertical dividers between rooms – the internal walls.

Stairs or lifts can provide access between storeys. Doors can provide access to the building and to each room within the building.

Daylight and ventilation can be introduced into the building by the provision of windows in the external walls or the roof.

All of these parts of the building construction are known as elements. Each of these elements will be considered individually within this book.

Self-assessment question 1.1

How would you define a building element?

Building loads

Buildings must not only be designed to support their own weight safely, they must also be able to support the weight of the occupants and any furniture, equipment or goods that are contained within the building. The building also needs to be able to withstand other forces that may be imposed upon it by the wind or snow.

Buildings are therefore designed to carry three specific types of loading:

- *Dead loading* – derived from the weight of the materials and components used in the building construction. This will alter according to the type of building being constructed.
- *Live loading* – derived from the weight of all the occupants using the building, together with the weight of all the furniture, equipment and goods that may be contained within the building. This will alter according to the use to which the building is put.
- *Superimposed loading* – derived from the forces exerted onto the building by the climatic conditions prevailing. This will alter according to the location of the building and whether it is exposed or sheltered.

Self-assessment question 1.2

Why is the weight of all the occupants using the building, along with the weight of all the furniture, equipment and goods, called live loading?

The total load that a building exerts on the ground beneath is obviously dependent on a number of factors, but as a rough guide, a typical brick built house will exert a total load of approximately 30kN/m (kilonewtons per metre run of loadbearing wall) on its foundations. This assumes that the external walls will carry the loading from the roof, the floors and the stairs directly to the foundations that are built beneath them. To ensure that the building will not suffer collapse if the design loads are exceeded during its lifetime, a *factor of safety* is normally added to the total load. This factor of safety is normally taken as two and has the effect of doubling the calculated load of the building in order to ensure that the foundations can be designed to be adequately strong to cope with occasional excessive loading.

This aspect will be looked at in more detail in Chapter 5 (foundations).

The functional requirements of a building

In order to be functional, buildings need to satisfy more than their primary function. Modern buildings have to satisfy a variety of other needs and these must be considered in the design of the elements and sub-elements that make up the building.

Sustainability

Sustainable construction covers numerous aspects of the building process. It starts with the design of the building, goes through the construction process and encompasses building performance issues throughout its use. The principles of sustainable construction are to:

- reduce the consumption of resources
- use recycleable resources where possible
- reduce wastage
- protect the environment
- reduce energy consumption
- reduce maintenance costs.

It is estimated that more than a quarter of the UK's carbon dioxide emissions come from the energy we use to heat, light and run our homes. The Code for Sustainable Homes, published by the Department for Communities and Local Government, measures the sustainability of a home against certain design categories and rates the whole home as a complete package, based on the Building Research Establishment Global's Ecohomes System. The design categories considered are:

- the amount of energy consumed; carbon dioxide emitted
- the amount of water used
- the amount of materials used in the construction
- the amount of surface water run-off
- the amount of waste generated in the construction process
- the amount of pollution generated by the construction process
- the health and well-being of the occupants
- the management of the construction process
- ecological considerations.

The Code uses a sustainability rating system from one to six stars, with six stars being awarded to the most sustainable homes. By meeting the requirements of the Code it is hoped that houses will have reduced greenhouse gas emissions to the environment, they will be better adapted to the effects of future climate change and they will have a reduced impact on the environment. In addition, running costs should be reduced and living and comfort standards improved. In addition the European Union have issued a Directive on the energy performance of buildings in the member states that requires all member states to provide legislation that requires all new homes to have an Energy Performance

Certificate, providing key information about the energy efficiency and carbon performance of the home. This will be considered further in the section on thermal insulation.

It is estimated that one-tenth of the UK's carbon dioxide emissions come from the production of building materials. Building materials and components should therefore be rated according to the following criteria:

- embodied energy content
- degree of recyclability
- degree of reclaimability
- degree of biodegradability.

The Building Research Establishment has produced a Green Guide to Specification that compares the environmental impact of different types of building specifications. It assesses materials and components used in various elements of the building across their life cycle against 13 environmental issues and rates them from A* to E, with A* having the least environmental impact.

The embodied energy content is the amount of energy used in the production and use of the material over its life cycle. This includes extraction, processing, manufacture, transportation, maintenance and disposal. The majority of this energy is consumed by a construction material up to the point that it is incorporated into the building.

The degree of recyclability incorporates the capability of the material to be recycled after its use in the building is over. It also includes any waste generated during the extraction, processing, production and installation processes that could be diverted back into the manufacture of new materials. The direct substitution of recycled materials for new in construction work can reduce the environmental impact of that material.

The degree of reclaimability is similar to that of recyclability. The difference is that with recycling, the waste material often requires reprocessing to produce a new material, whereas with reclamation, the materials are recovered from the demolition process, cleaned up, cut to the size required and re-used in their original form. It is estimated that up to 60 per cent of all demolition waste is now reclaimed in some form.

Self-assessment question 1.3

Give an example of a typical construction material that can have a high recycled content, and a typical construction material that can have a high reclaimability.

The degree of biodegradability depends to a great extent on the amount of organic matter that is contained within the material. Generally the greater the organic content, the greater the biodegradability. Alongside this the toxicity of the material also needs to be considered. Any materials going to landfill should have a high biodegradable content and have low or zero toxicity if they are not to pose a threat to the environment.

Self-assessment question 1.4

Give an example of a construction material that is highly biodegradable but has low toxicity.

The designer needs to consider sustainability when selecting the materials and methods of construction for the building. Materials requiring high energy consumption in their manufacture should be avoided as much as possible. However, the maintenance costs of the building during its lifetime should also be considered when initially selecting materials for its construction. Bricks need large amounts of energy in their manufacture, but can last a number of years without having to be replaced, so their *life cycle cost* (the cost of a particular element of a building calculated over the building's life; this will include the initial cost of the materials together with operational costs, maintenance costs and even replacement costs during the lifetime of the building) may be small compared with an alternative material such as softwood timber boarding or PVCu cladding panels.

Where possible, designers should try to use recyclable and reclaimed materials in construction. For instance, it is possible to use crushed concrete as an aggregate in new concrete, rather than extracting gravel or quarrying rock for aggregates. This is less harmful to the environment. Where possible, materials for building should be locally sourced, so reducing the transportation costs. Home-grown timber is a good example of a highly sustainable material. Production costs are low, transportation costs are low and the effect on the environment is minimal, as the felled trees are replaced with saplings. Timber is also a net consumer of CO_2, rather than a producer.

Construction techniques that should be considered at the design stage will include the amount of excavation required to construct the building. Excessive excavation will require large volumes of soil being exported to landfill, which can have harmful effects on the environment. Another effect on the environment can be the amount of waste generated by cut-offs from materials. If this waste cannot be recycled it will again go to fill up landfill sites. The designer can minimise this waste by considering the generic sizes of the materials and components being specified for the building and attempting to use these materials as near as possible to these sizes.

Design will also consider the reduction of energy and water consumption within the building. This will include making the maximum use of daylighting in the building to reduce the amount of artificial light required. The greatest amount of energy used within buildings is required to heat or cool them. The designer needs to consider how thermal insulation can be utilised most effectively to reduce heating and cooling costs and their effect on the environment. Many modern buildings are now incorporating renewable energy sources in their design. These may come from small wind turbines and photo-voltaic cells generating electricity for consumption within the building, and allowing any surplus to be sold to the electricity generating companies.

Heat pumps can extract heat from the air, ground or water and use this to heat the building. Solar panels mounted on the roof can extract energy from the sun and use this for heating water within the building. Water is also a valuable resource which needs to be carefully utilised within the building. Water wastage can be minimised by avoiding long pipe runs that require large draw-offs of water. Other techniques for reducing water wastage include recycling 'grey' water from baths and washbasins for use in flushing toilets. Collecting rainwater from roof run-offs can also be beneficial to the environment.

During the construction process, sustainability issues still need to be considered. The reduction of waste can be accomplished by good site organisation and management. The selection of contractors by clients who have strong environmental concerns will often be based not only on their construction cost, time and quality data but also by their environmental performance credentials. These can include the minimisation of damage and inconvenience to the surrounding environment.

The use of 'green' building techniques may often seem more expensive on an initial cost basis, but most of these techniques will prove economic and repay their initial cost with savings within a relatively short time period. Future costs of energy and water resources are bound to rise as these resources become scarcer and the payback period for these renewable energy and recycling systems will decrease.

Self-assessment question 1.5

An architect has included a large proportion of glazing in the design of a new house.

Consider the influence on the sustainability of the house on this design decision.

Strength and stability

If a building is to act as an enclosure, it must be strong enough to carry the loads imposed upon it and be stable enough to transfer those loads to the ground beneath safely. The strength of the building is related to the strength of the materials used in its construction.

As buildings are loaded they become subjected to a range of stresses that must be safely accommodated by the materials used in the structure. There are basically three kinds of stress that building materials will be subjected to:

- *compression* – pushing
- *tension* – pulling
- *shear* – tearing.

Self-assessment question 1.6

What are the stresses that are being put onto the following materials?

a) A brick wall supporting a roof.

b) A concrete floor with a heavy statue placed onto it.

c) A sheet fabric stretched over a timber framework.

A beam, such as a floor joist or roof rafter, spanning between two loadbearing supports will bend (or deflect) when loaded. This deflection is considered acceptable provided it does not cause the beam to fail (that is to crack or even to break) and provided the amount of bending is not unsightly or could cause finishes which are attached to the member to crack.

Different materials react in different ways under load. Some materials, like concrete, brick and glass, are very brittle and have low *ductility*. They will not bend very far before they begin to fail (usually because of tensile stresses). Other materials such as steel and timber have a higher ductility and can accommodate a greater amount of bending before they fail.

Some materials have a high *modulus of elasticity* and will bend very easily when loaded and return to their original shape when the load is removed. However, these materials are not as good for structural uses as materials that have greater stiffness.

In addition to loading considerations, building materials are also subjected to stresses that are caused by movement. This may be structural in nature, caused by the settling of the foundations or the bending or twisting caused by the application of a load, or could be caused by expansion and contraction of the material due to changes in its moisture content or temperature.

There may even be movements in the materials caused by chemical reactions between materials placed in contact with each other or with chemicals in the environment.

The building must therefore be designed with consideration being given to the structural properties of the materials being used. This is more important for loadbearing elements, such as walls, floors, roofs, foundations and stairs, than it is for non-loadbearing elements such as partitions, windows and doors. Another important consideration is the weight of the materials themselves, since heavy materials will add more to the total load of the building through their own dead weight than will lighter materials having similar strength characteristics. Thus an important ratio to be considered when selecting materials for structural applications is that of the material's strength to its weight. A high value is considered to be better than a low value for this ratio.

Self-assessment question 1.7

Which material is likely to have the better strength to weight ratio, concrete or timber?

Weather resistance

If a building is to provide shelter from the external climate then the parts of it that are exposed to that climate must be weather resistant. This is generally taken as being resistance to water penetration either from the rain through the external elements of the building (roofs, walls, windows and doors) or rising dampness from the ground through elements in contact with the ground (ground floors and footings to walls). However, air infiltration (draughts) may also be considered unacceptable to the occupants of the building and therefore may also need to be minimised.

Water penetration into the building may be reduced by four main mechanisms, either adopted singularly or in combination:

- Making the structure impermeable by using a barrier that will prevent the passage of water across it.
- Using specially shaped joints that will prevent the passage of water from the exterior to the interior of the building.
- Allowing a small amount of permeability due to the use of slightly porous materials, but designing the structure to be sufficiently thick enough to allow the path of water to be reversed, due to evaporation, following a change in the weather, before it reaches the interior of the building.
- Adopting a cavity, which water cannot cross and that separates the damp exterior of the building from the dry interior of the building.

This aspect will be looked at in more detail in Chapter 8 (external walls).

Self-assessment question 1.8

Which of the above techniques is used to provide weather resistance to a window?

In practice the latter two mechanisms tend to be more efficient than the first two, since most building materials have some permeability, and even if a material is totally impermeable to the penetration of water there can be problems at the joints or where small cracks have developed due to structural, thermal or moisture movement in the material or component.

This aspect will be looked at in more detail in Chapter 22 (external wall finishes).

Self-assessment question 1.9

Name two impermeable building materials.

However, one technique, which has been successfully used to combat water penetration seeping through a material, is to use an impermeable material placed in a strategic location to act as a barrier to damp penetration: the damp proof course. This is very effective at the base of walls in preventing moisture from the ground being conducted up through the permeable brickwork to the interior of the building, above ground level, by capillary action. This aspect will be looked at in more detail in Chapter 9 (bonding and opening in walls).

Self-assessment question 1.10

What is capillary action and how may it affect footings to walls?

Fire resistance

Buildings should be designed to be fire resistant and to withstand the effects of fire from within as well as from without, for sufficient time to allow for the escape of occupants to a place of safety.

It is preferable to construct buildings from materials that are non-combustible but this is not always feasible. Timber is a very versatile building material and yet it is combustible. However, if timber structural members such as beams, posts, joists or rafters are designed of sufficient thickness, then the rate of combustion of any excess or 'sacrificial' timber may provide sufficient time for the building to be safely evacuated before the structural member is sufficiently weakened to cause it to fail under load. Indeed, the charring of the 'sacrificial timber' around the member can actually slow down and, in some cases, even stop the rate of burning. This aspect is considered in more detail in Chapter 10 (timber).

Many structural building materials, like brick, have good fire resistance since they have been produced through the refractory process, in a kiln, and are therefore less susceptible to combustibility or loss of strength at high temperatures than other materials. Steel, however, although it is produced in a furnace, can suffer rapid loss of strength once its temperature exceeds 500°C (quite possible in a well-established fire) and is likely to buckle under the stresses imparted by the loads the member may be carrying. Concrete can also suffer damage in a fire if natural aggregates are used. The moisture present in the particles of coarse aggregate may expand as it is heated beyond boiling point causing the particles to explode and the concrete to *sinter*.

Even non-structural materials may need their fire resistance to be considered in the design of a building. Finishes to walls, floors and ceilings may allow the surface spread of flame, thus causing the fire to spread rapidly inside a building. Materials with a low surface spread of flame value may need to be considered in these applications. This aspect is considered in more detail in Chapter 21 (internal finishes).

British Standard 476 (BS 476) Fire tests on building materials and structures outlines three main fire tests on building materials to determine their performance in a fire. These are:

- *Stability* – freedom from collapse due to the action of fire
- *Integrity* – resistance to the passage of flame and smoke

- *Insulation* – resistance to the passage of heat that may cause spontaneous combustion of other materials not in direct contact with the fire.

This aspect will be considered further in Chapter 17 (doors).

Components may be protected from the effects of fire by retardant treatments or by covering them with materials having high fire resistance.

Self-assessment question 1.11

Give an example of a material that has high fire resistance.

Thermal insulation

Modern buildings need to be designed to satisfy the demanding comfort conditions of occupants. A primary consideration in this respect is thermal comfort. This is related to the capacity of the materials within the building to store heat. It is also affected by the external air temperature, the amount of heat gained from the sun, the amount of ventilation (draughts can make a room very uncomfortable), the air tightness of the structure and the amount of thermal insulation provided in the building.

Buildings need to be insulated to prevent excessive amounts of heat escaping to the outside, and may also be required to prevent excessive heat or cold on the outside from affecting the internal environment.

Because of the high cost of energy and the climatic conditions in this country, much attention has been paid in recent years to retaining heat within the building. Materials having a high *thermal mass* are good at retaining heat but they take longer to heat up than buildings with a low thermal mass. Materials having a high thermal mass tend to be dense and have a low *thermal conductivity* (it takes longer for heat to be conducted through them).

Self-assessment question 1.12

Give an example of a material having a high thermal mass.

In other countries, where climatic conditions are different to our own, more attention is paid to preventing excessive heat from the outside from entering the building. Both circumstances can be improved by the introduction of thermal insulation.

Air is an extremely efficient *thermal insulant*; therefore most thermal insulation materials incorporate a large amount of air voids. Some building materials are naturally good thermal insulants, whilst others are notoriously poor and therefore need the addition of materials with a high *thermal resistivity* value (a good resistance to the flow of heat through them) to improve their thermal performance. Thermal resistivity is the inverse of thermal conductivity. Most manufacturers of building materials now quote the thermal conductivity value in W/m²K (Watts per square metre per degree Kelvin) rather than their resistivity value.

Self-assessment question 1.13

Give an example of a building material that has a high thermal resistivity value and one that has a low resistivity value.

Materials used to add thermal insulation to a building can be classified into three categories:

- plant and animal fibre-based insulants
- cellular plastic insulants
- mineral, stone or glass fibre insulants.

Plant and animal fibre-based insulants are organic, such as cellulose fibre (often made from pulped, recycled paper) and sheep's wool, have very good sustainability credentials and typically have a thermal conductivity value of 0.038W/m²K (the resistance of the material to heat flowing through it is generally measured by the rate at which heat is conducted through the material measured in Watts per square metre of material, per degree Kelvin (the lower the thermal conductivity, the better the material as an insulator)).

Cellular plastic insulants are inorganic materials and generally have poor biodegradability. However, they have low thermal conductivity levels (typically around 0.030W/m²K) and are therefore considered to be the best insulators. They are also available as rigid slabs which makes them easier to install in vertical structures, such as walls, than fibrous materials that can slump. They are also more advantageous in areas such as flat roofs that may be subject to foot traffic (such as balconies or where access for maintenance may require walking across the roof). This is covered in more detail in Chapters 8 and 13 (external walls; flat roof construction and coverings).

Mineral fibre insulants include stone wool and glass fibre. Stone wool is produced by heating a volcanic rock to 1600°C in a furnace, adding a binder to cement the fibres together and then spinning the fibres of molten rock in a similar manner to candy floss. Glass fibre is made in a similar manner using recycled glass, heated to 1450°C and combined with a binder. Both materials have good fire resistance properties. They rely on the air trapped between the fibres to provide good insulation, with thermal conductivity levels typically around 0.035W/m²K. Although they are supplied either on a roll for ease of application in a loft space or as more compact slab materials, like plant and animal fibre-based insulants, they are easily compressed if a load is applied to them, and once the trapped air has been forced out by compression their thermal insulation performance is reduced.

Of course there can be problems with trying to improve the thermal resistance of building materials, since large amounts of air voids in an external element may affect its weather resistance.

Self-assessment question 1.14

Why may large amounts of air voids in an external element affect its weather resistance?

For many years building control legislation within the UK has concentrated on the improvement of thermal comfort and the reduction of heat loss in houses by stipulating the amount of thermal insulation that should be provided within the construction of walls, roofs and ground floors. A method of measuring the thermal performance of these elements was devised by totalling up the *thermal resistances* of all the component parts of the element over a square metre area per degree drop in temperature either side of the element in question. Thermal resistance is not the same as thermal resistivity, since it includes the thickness of the material being considered. Thus a brick wall constructed of bricks having the same thermal resistivity value would have a different thermal resistance value for a 1B thick wall to that of a 1½B thick wall.

Self-assessment question 1.15

A 150mm thick aerated concrete block has a resistivity value of 9.09W/m²K. Calculate its thermal resistance value and thermal conductivity value.

To this value was added the surface resistances of the inside and outside of the element being measured. The reciprocal of the sum total of resistances was then taken to provide the overall heat transfer coefficient or *U-value* measured in W/m²K. A low U-value denotes low heat transfer through the structure and therefore a high thermal efficiency. To account for the different methods of construction of each element an area weighted average is assumed. It is this U-value that has been used in the legislation (called the Building Regulations) to provide a minimum target for the thermal performance of external walls, roofs and ground floors that must be achieved in their construction.

Self-assessment question 1.16

A typical external wall has an external surface resistance value of 0.04, a ½B thick external leaf wall 102.5mm in thickness constructed of bricks having a thermal conductivity of 0.77, a 50mm unventilated air cavity having a thermal resistance of 0.18, a layer of rigid polyurethane thermal insulation 50mm thick having a thermal conductivity of 0.03, a 150mm thick aerated concrete block internal leaf wall having a thermal conductivity of 0.11, a 12.5mm thick plasterboard internal wall lining having a thermal conductivity of 0.16

and an internal surface resistance of 0.13. Calculate its U-value (all values are in W/m^2K).

Hint: First calculate the thermal resistance value for each material, total these up and then add the external surface resistance value, the air cavity thermal resistance value and the internal surface resistance value, then take the reciprocal of the total to obtain the U-value.

The calculation of U-values for different elements of the structure are in fact more complicated than is shown above. For instance, the thermal conductivity values for the wall ties and the mortar in the brick wall and the block wall need to be included. This will slightly raise the U-value for the wall.

The development of thermal insulants has now reached a point at which further reductions in the thermal performance of these elements can only be achieved by increasing the thickness of the thermal insulation being used (thus increasing their thermal resistance value). This is likely to make the building elements very thick and may also jeopardise their structural integrity in terms of how they are able to support or carry the building loads imposed upon them. This will be discussed in more detail in Chapter 8 (external walls). Therefore the Building Regulations 2010 imposed further requirements on improving the thermal performance of buildings in an effort to reduce the amount of CO_2 emitted from a building by 20 per cent. This was in response to the EU Directive on the energy performance of homes. In order for new houses to achieve an Energy Performance Certificate, each building is given a Standard Assessment Procedure (SAP) rating on a scale of 1–120 (where a score of 100 indicates no heating or hot water cost and ratings above this level indicate generation of electricity exported to the national grid) based on its annual energy costs for space and water heating. From this a Carbon Index (CI) measures the amount of CO_2 emissions from the building associated with space and water heating on a scale of 0–10. The CO_2 emission of a notional dwelling built to 2002 standards is calculated using the Standard Assessment Procedure. This depends on a number of factors such as:

- the materials used in the construction of the building elements
- the thermal mass of the building fabric
- the amount of thermal insulation used within the building fabric
- the ventilation characteristics of the building
- the type of heating system, its efficiency and controls
- the amount of solar heat gain to the building
- the type and quantity of fuel used for space and water heating, lighting and any mechanical ventilation
- the use of renewable energy systems.

This is then reduced by 20 per cent to give a Target Emission Rate (TER) for the dwelling to be constructed. The designer must then calculate the actual Dwelling Emission Rate (DER), which must not be worse than the Target Emission Rate. SAP, TER and DER ratings calculations can be quite complex, so computer software programs have been developed to make this easier.

The Building Regulations Approved Document L1A Conservation of fuel and power outlines the techniques for reducing the CO_2 emissions of new houses by 20 per cent. As well as providing sufficient thermal insulation to building elements, it also makes provision for:

- improving the thermal efficiency of the walls, windows and roof by using more insulation or better glass
- reducing air permeability to the minimum consistent with health requirements
- carefully designing the fabric of the house to reduce thermal bridging
- limiting the effects of solar gains
- installing a high efficiency condensing boiler
- encouraging the use of solar thermal panels, biomass boilers, wind turbines and combined heat and power systems to replace energy taken from the National Grid with low or zero rated carbon generated energy.

It is estimated that air leakage accounts for 5–10 per cent of heat loss within a house as well as draughts affecting thermal comfort. Draughtproofing to windows, external doors and loft hatches needs to be improved. In addition, linings to external walls, joints in vapour control layers and breather membranes need to be effectively sealed to reduce air leakage. This is considered in more detail in Chapter 15 (timber frame and steel frame house construction). The *air permeability* of the dwelling is measured as the amount of air leakage per hour per square metre of the building envelope at a test reference pressure of 50 Pascal ($50N/m^2$) differential across the building envelope. This should not be greater than $10m^3/h/m^2$. Obviously there has to be some air leakage within the building to ensure adequate natural ventilation, but the object of these requirements is to reduce excessive air leakage and its consequent effect on heat loss from the building.

Thermal bridging can be defined as the introduction of a material having a high thermal conductivity value which bypasses the thermal insulation material in a building element. It is estimated that 10–15 per cent of heat losses from houses can be attributed to thermal bridging. Clearly, heat losses can occur through timber or steel studs in external wall construction, timber or steel joists in floor constructions and timber rafters in roof construction where these materials continue beyond the insulation layer. In addition, heat can be lost through thermal bridges created at the junctions of floors and roofs with the external wall and these junctions need careful detailing. This is assessed in the *Building Research Establishment Information Paper IP 1/06 Assessing the effects of thermal bridging at junctions and around openings*.

Provisions also need to be made to limit the rise in internal temperatures due to effects of solar gains. This can be achieved by carefully designing the size, amount and orientation of the windows in the house, together with protection by the use of shading and ventilation. This is considered in more detail in Chapter 16 (windows and glazing).

To assist the designer in achieving the standards required by the Building Regulations Approved Document L1A, a set of Accredited Construction Details have been prepared by the Department of Communities and Local Government.

Sound insulation

Modern buildings also require adequate levels of acoustic insulation to prevent external noise from affecting the internal environment or noise created within the building from affecting the external environment.

Sound is transmitted in buildings by two main mechanisms:

- *Airborne transmission* – sound waves caused by speech or music travel through the air and impinge onto elements of the structure (such as walls) and cause them to vibrate as a diaphragm in sympathy with the wave pattern, thus transmitting the sound wave to the air on the other side of the element.
- *Structure borne or impact transmission* – impacts caused by hammering or footsteps cause the molecules in the element of the structure (such as floors) to vibrate in sympathy. This energy is then passed on to other molecules in adjoining elements, causing the sound to be transmitted to other parts of the building.

Insulation against sound transmission is different for the two mechanisms of transmission. The best insulator against airborne transmission is mass, since bulky elements are less likely to vibrate in sympathy with the sound waves than are flimsy structures. Insulation against structure borne or impact transmission is discontinuity, since it is more difficult for the sound energy to be passed from the molecules in one element to those in an adjoining element if they are separated from each other. This may be difficult to achieve if one element relies on the other for structural support.

A further improvement in sound insulation can be accomplished by reducing the amount of gaps or open pores in the surface of the material. Thus joints between materials or components may need to be effectively sealed, as may their surfaces.

Self-assessment question 1.17

How can the open pores in the surface of a wall constructed from concrete blockwork be effectively sealed to improve its sound insulation?

Durability

Buildings do not last forever, but are extremely expensive to construct and maintain. It is therefore important to ensure that sufficient thought is given to the selection of materials and components used in the construction of buildings regarding their durability and ease of maintenance and repair, particularly in those areas where replacement of the material or component during the life of the building is likely to be extremely difficult.

The durability of a building material may be related to its ease or frequency of maintenance, repair or replacement. The location of the material or component may also have an effect on its required durability.

If the material or component is exposed to the external environment then frequent wetting and drying or changes in temperature between night and day may have an effect on its performance. Absorbent materials may be subjected to frost damage when the water they have absorbed freezes at sub-zero temperatures and the ice lenses expand within the material. Clay products may contain soluble salts which may re-dissolve in water taken up by absorption but be left on the surface of the material as efflorescence, when the absorbed water later evaporates. This aspect will be considered more fully in Chapter 7 (bricks and blocks).

Roofing felts and some plastics can become brittle and crack due to prolonged exposure to ultra-violet rays from the sun. This aspect will be looked at again in Chapters 13 (flat roof construction and coverings) and 16 (windows and glazing).

Materials and components that form essential parts of the structural fabric of the building cannot easily be replaced and consideration therefore needs to be given to their designed life expectancy that should match that of the building itself. Internally, the durability and maintenance of finishes may have more to do with the ease and frequency of cleaning as well as the material's resistance to indentation or impact caused by the wear and tear of everyday use. Clearly, wear and tear will be directly related to the proposed use of the building. For instance, a wall finish in a school will probably get more abuse than the same finish in a house. This aspect will be looked at in more detail in Chapter 21 (internal finishes).

The cost of maintenance, repair or replacement of materials and components needs to be considered alongside their initial cost when specifying their inclusion in a building. The concept of *life cycle costing* considers the overall cost of using that particular material or component during the life of the building. When maintenance costs are considered it may be found that a material which has a high initial cost may have a lower life cycle cost (because it needs less maintenance during its life) than an equivalent material which has a much lower initial cost but high maintenance cost.

This aspect will be looked at in more detail in Chapter 17 (doors).

Appearance

An important consideration in any design is how the product looks, and buildings are no exception. Careful thought therefore needs to be given to the appearance of the materials and components used in the building and how that appearance may be affected by exposure to the effects of the weather.

The importance of the appearance of a particular building material or component will largely depend on whether it can be seen or not. Finishes may cover the structural elements of a building and it is therefore these that will have their aesthetic appeal considered in their selection. This aspect will be considered further in Chapters 21 and 22 (internal finishes; external wall finishes).

The architectural design of the building will also play an important part in its appearance. Appearance can also be affected by the use of decorative coatings and these are widely used on many synthetic materials. Natural materials such as bricks and stones may use texturing to create a different appearance.

Quality

Finally, quality is also an important consideration in the design and manufacture of any product. If buildings are to have a long working life then they must be capable of fulfilling the needs of their users as effectively as possible. It is generally accepted that the quality of a commodity is directly related to its price, and this relationship is true of buildings also. We accept with other commodities that we try to obtain the best quality that we can afford and we must be prepared to relate this also to the acquisition of buildings. Quality is a difficult concept to adequately define and is best considered as 'fitness for purpose', which attempts to relate the quality of the product to its intended use.

Although not strictly a functional requirement, quality is still regarded as an important consideration in the selection and use of building materials and components. It embodies the performance of the material or component against every other functional criterion. It is extremely difficult to determine whether a material that has better weather resistance than fire resistance is better quality than a material that has better thermal insulation than sound insulation. Each of the functional requirements considered above need to be considered separately.

Consideration has already been given to the problem of a material with a high thermal performance having poor weather resistance and these trade-offs may need to be made for other situations. It is therefore important that not only are the functional requirements of a building element considered in the selection of an appropriate material or component, but also their relative significance. It will be necessary to prioritise the functional criteria in order of importance and this may involve a system of weighing these criteria against each other.

It has already been stated that there is a strong correlation between quality and cost. The question may need to be asked, 'What standard of quality can I afford?' In the final analysis does a building client require the same level of quality for the ground floor of a warehouse as they do for the ground floor of a five star hotel?

Self-assessment question 1.18

Give an example of a building element where its sound insulation may be more important than its thermal insulation and one in which its weather resistance may be more important than its fire resistance.

 Visit the companion website to test your understanding of Chapter 1 with a multiple choice questionnaire.

CHAPTER 2

Development controls

Introduction

Before most types of building development can commence, whether residential, industrial or commercial, there are a number of statutory requirements that must be met. These are laws that are passed by the government to ensure that any development is controlled within acceptable limits. The main controls affecting building development are Planning, Building Control, and Health and Safety. Planning controls consider where a building is going to be placed and what it will look like; Building Control focuses on how the building is constructed, and Health and Safety considers the safe construction, use and disposal of the building. This chapter will briefly consider these three development controls, how they operate and what a developer must do to comply with their requirements.

Planning

Planning, sometimes also referred to as Town and Country Planning, has been used to control development from early times. One has only to look at Roman settlements to see that thoroughfares were laid out to a pre-determined pattern and municipal buildings were situated in particular locations. Without Planning Controls development would spread in a haphazard fashion and one building activity could encroach on another in an unacceptable manner. For instance, it is unlikely that anyone would want their house to be situated next to a noisy factory or for factories and warehouses to be built where there is poor transport access to them. It is also important to ensure that cities and towns do not spread out and spoil the surrounding countryside. Where development of new buildings is acceptable, controls need to be exercised as to how the building will look. Will it fit in with the surrounding buildings? Will there be sufficient roads, drainage, parking spaces and water, gas, electricity and telecommunications services (the infrastructure) to ensure that the function of the building is served successfully?

Planning is ostensibly controlled at a national level by the government through legislation; this mainly takes the form of Acts of Parliament and

Statutory Instruments controlled through the Department of Communities and Local Government. The latest items of legislation with regard to planning are the Planning Act 2008 and the Housing Act 2004. These provide planning guidance. Enforcement of the legislation is devolved to local authorities. The planning system requires each local authority to provide a local planning authority, which prepares a Local Development Framework. This outlines how planning will be managed in that area. As part of this framework the local planning authority will also produce Development Plan Documents. These contain the statutory policies and proposals that guide development in the local authority area and will contain the core planning strategy, the regional spatial strategy, the sustainable community strategy and a statement of community involvement in its preparation. Basically each authority needs to review the land contained within its area and consider its current use and future use over a period of 5–10 years. The land can be classified into categories of use such as agricultural, residential, industrial, commercial and recreational and the infrastructure (roads and services) needed to serve this proposed land use also needs to be considered. The future use of the land will be based on forecasts of how the area is likely to develop and must take into account the needs of the communities involved. Once the Development Plan Documents have been prepared, they are submitted for inspection by independent planning inspectors. If they are passed then the Development Plan Documents become the basis for deciding on planning applications.

Most building work requires planning permission; however certain loft conversions, extensions, conservatories, porches and outbuildings to existing houses may come under permitted development and therefore might not require planning permission. When the building of a new house is being considered, the developer may apply for either outline planning permission or full planning permission. Outline planning permission is generally used to find out, at an early stage, whether or not a proposal is likely to be approved by the planning authority, before any substantial costs are incurred. It allows for fewer details about the proposal to be submitted and may just include a location plan showing the site area and surrounds along with a brief description of the proposed development. The local planning authority will consider this application and may then pass the proposal with a requirement for Reserved Matters to be considered at a later stage. This will consider such aspects as appearance (the way the building looks), the layout (the way routes and open spaces within the development are laid out), scale (the size of the development, including the height, width and length of each proposed building) and landscaping (including the planting of trees and hedges to screen the development). These will need to be submitted when the developer applies for full planning permission.

Self-assessment question 2.1

Why may a developer wish to apply for outline planning approval first rather than go for full planning approval?

Alternatively the developer may decide to apply for full planning permission straightaway and will be required to supply a location plan, a site plan (providing details of the proposed development), floor plans and elevations of all buildings on the proposed development, an ownership certificate (showing that either the applicant is the owner of the land or is applying on behalf of the owner) and a Design and Access Statement. This statement outlines the design principles that have been applied to the proposed development and how issues relating to access to the development have been dealt with.

Self-assessment question 2.2

Why does the local planning authority require an ownership certificate showing that the applicant is either the owner of the land or is applying on behalf of the owner?

Qualified planning officers will look at the proposal and how it relates to the Local Development Framework and in particular the Local Development Plan, and seek comments from the local community regarding any objections to the proposal. They will then prepare a report on the application which will be submitted to the planning committee of the local authority. This committee comprises elected councillors who have an interest in local development issues but have no professional experience regarding compliance with planning legislation. They are guided in their decision making by the report submitted by the planning officers. The decision taken by the planning committee will determine whether an application is accepted or rejected. There is an appeals process, firstly through the local planning authority and finally, if required, through the Department of Communities and Local Government. Upon approval a development must be commenced within three years otherwise the approval lapses and the application process must commence again.

If a developer does not apply for planning permission and the proposed development does not fall within the rules for permitted development, or the development does not comply with the proposal that was approved, the local planning authority may issue a demolition order. In many cases, however, a developer may apply for retrospective planning permission.

Building Control

Controls on the design and construction of buildings to ensure the safety and health of people using the buildings have also been in force for many years. Regulations on building in London are believed to date back to 1189 and were followed by other cities in later years, but were poorly enforced. This was amply illustrated by the Great Fire of London in 1666, which led to the passing of the London Building Act in 1667, in which surveyors were given powers to enforce compliance with the regulations. Gradually other cities and towns throughout England and Wales adopted similar principles but it was not until the Local Government Act of 1858 that regulation of the structure of buildings through local bye-laws was introduced. These bye-laws varied from one local authority to another and could be confusing to developers who were building in a number of different areas with differing building bye-laws applying. An attempt to produce model bye-laws was introduced in the Public Health Act of 1936, but these were simply guidelines, and requirements could still vary from one authority to another. Scotland was the first country in the United Kingdom to adopt national regulations with the passing of the Building (Scotland) Act in 1959. In England and Wales the Public Health Act of 1961 created the first national Building Regulations (apart from the Inner London Boroughs, Scotland and Northern Ireland) which were published in 1965. The Building Regulations (Northern Ireland) Order of 1972 established regulations for Northern Ireland modelled closely on those for England and Wales. There have been several subsequent revisions of the regulations, the most significant of which was brought in by the Building Act 1984. The Building Regulations 1985 provided a more flexible approach to the design of buildings by cataloguing the basic functional requirements for buildings in a set of Approved Documents. The Building Act 1984 also allowed for the creation of approved inspectors, who could operate the building control function as an alternative to local authority building control officers. In recent years the Building Regulations have gone further than ensuring the safety and health of people using buildings. They now set high standards for the conservation of heat and power and also provide for access and use of buildings and facilities for the disabled.

On obtaining planning consent, developers then need to apply for building control approval for most types of building work. Drawings that have previously been passed for approval in the planning permission process will most likely not be sufficient for seeking approval through the building control process. Planning drawings focus on the aesthetics of a project and how the development will fit in with its surroundings. Drawings used to seek approval from building control are more technical, detailing how individual sections of the building are to be constructed and are often accompanied by a specification that identifies sizes and quantities of individual elements of the building.

Not all building projects require formal approval or a Building Regulations application. Works that may be exempt include porches, conservatories, detached garages, carports, greenhouses, sheds and other temporary buildings, along with like for like replacement and repair of existing drainage. Operation of the building control process is devolved by the Department of Communities and Local Government to local authorities or approved inspectors. Persons or organisations wishing to become approved inspectors need to apply for registration to the Construction Industry Council. They need to demonstrate a comprehensive knowledge base of Building Regulations and statutory control, building law, construction technology and materials, fire studies, foundation and structural engineering and building services and environmental engineering. They are then interviewed to assess their level of competence and must obtain professional indemnity insurance. If they are accepted, registration lasts for five years, after which time they must re-apply.

Self-assessment question 2.3

Under what circumstances would a developer find it more useful to use an approved inspector than a local authority building control officer?

Applicants for building control may apply for full plans approval or a building notice. A full plans application has to be accompanied by detailed drawings. The building control officer or approved inspector then checks the plans for compliance with the requirements of the Building Regulations as set out in the guidance provided in the Approved Documents. A full plans application may be approved with conditions attached, which highlights where the application does not meet the regulations. This allows the project to proceed and the conditions are then discharged during the building process, when the regulations have been satisfied. A building notice application is generally made for minor domestic works, generally involving improvements. It does not require plans to be submitted for inspection and work can begin 48 hours after the application has been received. A building notice cannot be used for industrial or commercial developments.

The Approved Documents provide practical guidance as to how the requirements of the Building Regulations may be met. They do this by outlining methods of construction that meet the regulations. These Approved Documents are separated into 14 parts:

Part A – Structure
Part B – Fire safety
Part C – Site preparation and resistance to contaminants and moisture
Part D – Toxic substances
Part E – Resistance to the passage of sound
Part F – Ventilation
Part G – Sanitation, hot water safety and water efficiency
Part H – Drainage and water disposal
Part J – Combustion appliances and fuel storage
Part K – Protection from falling, collision and impact
Part L – Conservation of fuel and power
Part M – Access to and use of buildings
Part N – Glazing: safety in relation to impact, opening and closing
Part P – Electrical safety: dwellings

These Approved Documents are reviewed and updated periodically so as to enable continual improvement of building elements to occur, as required by government standards. However, these examples are not the only way in which compliance with the regulations can be met. Applicants may suggest other ways in which their proposed design will meet the requirements and it is the responsibility of the building control officer or the approved inspector to check that these proposals are acceptable. Occasionally they may ask for more information to support an application, such as a more detailed drawing of a particular aspect of the works or structural calculations for determining the size of a supporting beam or foundations.

If the plans are accepted then the applicant may start work on the building, but will be required to notify the building control officer or approved inspector when certain stages of the building process have been completed, so that they may be inspected.

Self-assessment question 2.4

Why should the building control officer need to inspect the work when certain stages of the building process have been completed?

Approved inspectors can inspect works but do not have enforcement powers where non-compliance is detected. In this case they must inform the local authority building control department, who will take over the task of inspection and compliance. Any work that does not comply with the requirements of the Building Regulations will need to be rectified. At the end of the project the developer will be issued with a final certificate stating that the work is compliant with the Building Regulations requirements. The lack of a completion certificate can affect the developer's ability to sell a property. If work has been carried out without a Building Regulations application, a Regularisation Certificate can be applied for (this applies to work carried out after 11 November 1985). A regularisation application can result in extensive work having to be undertaken in order to bring the building back up to standard. Full details and plans showing what has been carried out will need to be provided to building control. Works may have to be opened or uncovered to show compliance with the Building Regulations before a Regularisation Certificate will be issued.

Health and Safety

Construction is regarded as a high risk industry by the Health and Safety Executive. It reports that the construction industry accounts for 27 per cent of all fatal injuries to employees in the UK and 9 per cent of all reported major injuries, even though it employs only 5 per cent of all employees. It is therefore imperative that health and safety within the industry needs to be managed effectively.

The health and safety requirements of a construction project are covered by the Construction Design and Management (CDM) Regulations, administered by the Health and Safety Executive (HSE). The CDM Regulations have been designed to integrate health and safety into the management of a construction project from the design stage until completion by identifying risks early so that they may be reduced or eliminated at the design stage and ensuring that the remaining risks are properly managed during the construction phase of the project. Furthermore they consider the management of hazards and risks associated with the building throughout its lifetime and up to the point of its demolition and disposal. The aim is to make health and safety management an essential and normal aspect of construction projects.

The CDM Regulations apply to all construction projects, but a distinction is drawn between large and small projects, based on the number of anticipated working days that the project will require. Projects lasting more than 30 days or 500 working man days, whichever is the least, are classified as notifiable under the regulations, meaning that more stringent controls are required to be put into place and the results notified to the Health and Safety Executive.

Within the regulations four main parties are identified, and their duties and responsibilities are defined. These are:

- the client
- the CDM co-ordinator
- the designer
- the principal contractor.

The CDM Regulations also distinguish between domestic clients and others. A domestic client is someone who has work done on their own home but is not engaged in the trade or business of a property developer. It is likely that a project to build a new house would take longer than 30 days and so would be classified as a notifiable project. Much house construction in the UK is undertaken by speculative house-building companies and, for the purposes of the regulations, these are classified as clients, but they may also be the designer, the principal contractor and even the CDM co-ordinator.

The client must plan and allow sufficient time for all stages of the project. Where designers require additional information such as topographical surveys or soil investigations, then it is the client's duty to commission these and to ensure the information is passed to the design and construction teams. The client also has a duty to ensure that effective health and safety management is in place prior to the commencement of the project and throughout the duration of the project.

The client must also ensure that all appointees are competent to carry out their duties. On notifiable projects, the client is required to appoint a CDM co-ordinator to act as the key health and safety adviser on the project. The client will rely on advice given by this co-ordinator and the designer throughout the project. The client also needs to appoint a principal contractor, who will be required to oversee the health and safety requirements during the construction phase of the project. The client must also ensure that a health and safety file for the project is prepared and kept up to date. This should contain the information about the project which will enable future construction work, including cleaning maintenance, alterations, refurbishment and demolition, to be carried out safely. It should identify hazards and risks so that operatives involved in these subsequent activities can be aware of them and make appropriate arrangements to manage them. It should be kept up to date during the duration of the project and then stored and kept up to date, as necessary, throughout the lifetime of the structure.

The CDM co-ordinator is the key adviser in respect of health and safety risk management. He/she is responsible for notifying the HSE of information at important stages during the project. These are set out in Schedule 1 of the CDM Regulations. The co-ordinator must ensure that the client provides the necessary pre-construction information required by the design team and that the design team and principal contractor are complying with their duties under the regulations. He/she needs to facilitate good communication and co-operation between all parties involved in the project. The co-ordinator is also required to assist the client in selecting the principal contractor based on their competence to undertake the duties required of them under the regulations and ensuring that their management arrangements for health and safety on the project are adequate. The co-ordinator has the additional responsibility of compiling the health and safety file. The co-ordinator can be appointed as an independent adviser or can take on the role alongside the duties of any other member of the project team.

Self-assessment question 2.5

Why might it be advisable for the CDM co-ordinator to be an independent member of the project team?

The designer has a duty to ensure that as far as practicably possible the project is built safely and can be used, maintained and eventually demolished safely. Hazards need to be identified and eliminated or, where this is not possible, reduced. Where a hazard remains, the designer must notify the people likely to be affected by its existence. On notifiable projects the designer is required to acquaint the client with his/her duties, to ensure that a CDM co-ordinator is appointed and that the HSE have been notified about the project. A CDM co-ordinator must be appointed before detailed design work commences. The designer is also responsible, where required, to supply information to the co-ordinator for inclusion in the health and safety file.

The principal contractor should be appointed to the project as early as possible as he/she can provide valuable advice to the designer, particularly where complex work containing hazards or where high risk work is involved. As part of the health and safety management on site it is the principal contractor's responsibility to provide a construction phase health and safety plan to identify the hazards and assess the risks associated with their work and to supervise and monitor the work on site to ensure it is accomplished safely.

The principal contractor needs to check the competence of all employees working on the site, together with the competence of all subcontractors working on the site, and to ensure that all employees are supplied with all relevant information regarding health and safety and that they are trained accordingly. The principal contractor will need to liaise closely with all subcontractors to establish the preparation time they require prior to them commencing on site and to ensure that all relevant parts of the construction phase health and safety plan are provided to them. It is also the principal contractor's responsibility to ensure that all subcontractors have provided adequate health and safety training to all their employees on the site and that suitable welfare facilities are provided for all personnel on the site. All personnel should also be given site safety inductions before commencing work on site. The principal contractor is also responsible, where required, to supply information to the co-ordinator for inclusion in the health and safety file.

In addition to these four key parties the CDM Regulations also consider two further parties who have a role to play in managing health and safety on a construction project. These are the subcontractors (CDM calls them the contractors – this can sometimes lead to confusion as they may be considered to be the same as the principal contractor) and the workers.

Self-assessment question 2.6

Why should the duties of the principal contractor under the CDM Regulations be different from those of the contractor?

Subcontractors are required to plan, manage and monitor their own work on the project and that of their workers. In addition they need to check the competence of all their employees, train them in health and safety matters where required and supply them with relevant information concerning health and safety. They also need to ensure that there are adequate welfare facilities for all their personnel on site. On notifiable contracts they should contact the CDM co-ordinator when they commence work on the project, and liaise closely with the principal contractor about their work on the project, and co-operate with them on planning and managing the work, including informing the principal contractor of any problems they are aware of with the health and safety plan. They are also expected to contribute all necessary information to be included within the health and safety file and inform the principal contractor of any accidents, diseases and dangerous occurrences that they may be aware of, related to their work on site.

All workers on construction projects need to make themselves aware of their own competence and safe working practices and not work in an unsafe manner. If they spot a risk or dangerous occurrence then they must inform the subcontractor or the principal contractor. They should also ensure that the work they undertake does not compromise the health and safety of others on the site, and follow the health and safety procedures laid down in the health and safety plan for the project.

 Visit the companion website to test your understanding of Chapter 2 with a multiple choice questionnaire.

CHAPTER 3

Site investigation and the type of ground

Introduction

In this chapter it is intended to consider the investigations that need to be undertaken before selecting a site to build upon. Part of these investigations will consider the ground on which the building is to be constructed. The characteristics of the main types of ground commonly found in excavations on housing sites will therefore be examined and how these characteristics affect the behaviour of the ground, particularly when it is loaded by a building, will be considered. Because it is necessary to test the soil in order to determine its nature and likely characteristics, a small section of this chapter will be devoted to soil sampling and testing techniques.

It is not intended to go into significant detail about the principles of soil mechanics or ground engineering. The intention is merely to introduce the main types of ground likely to be encountered on construction sites, to consider the main characteristics of each type and to attempt to explain how these characteristics can affect the behaviour of the ground, especially under loading conditions.

Site investigations

Following the identification of a potential site for building upon, it is important to conduct a site investigation to determine whether the proposed land is suitable for its intended purpose.

There are two main forms of investigation:

- the desk study
- the physical or site study.

The desk study

Before even venturing onto the site it is possible to find out a great deal of information about the land being considered. From ordnance survey maps it is possible to determine the location of the site and the proximity of transportation links (road, rail, air and water). It is also possible to ascertain the level of the land and the levels of the surrounding land. By consulting geological maps of the area it is possible to determine the predominant type of soil or rock for the land being considered and the proximity of water courses.

Following this, it is possible to ascertain restrictions that may apply to the land, such as Town and Country Planning restrictions (e.g. is the proposed site in an area where an application for residential building is likely to be acceptable to the local planning authority or will it be rejected, perhaps because it falls within a 'green belt' area?). There may be rights of light, rights of support or rights of way that have been imposed on the land over the years and which must be maintained. There may be tunnels, mine workings (either active or abandoned) under the land. There may be ancient monuments or other archaeological interests on the land that may curtail its development.

Self-assessment question 3.1

Why should the possible location of tunnels or mine workings below the land be of interest in a site investigation?

The past history of the land being considered can also be investigated. This will be useful in order to ascertain whether the land has previously been developed and if so what kind of development there was. Some industrial processes could have left the land contaminated and this would need to be cleaned before the land could be developed for the construction of residential properties. The past history of the land may also establish whether the land has ever been used for landfill operations.

Self-assessment question 3.2

Why should it be important to clean land that has previously been used for industrial processes before re-using it for residential development?

It will be possible to establish whether any water, electricity, gas, telephone, sewerage and other services are connected to the land. If they are not, it will be important to determine whether it is possible to connect these services to any proposed building on the land and at what cost.

It will then be necessary to establish more about the area in which the proposed site is located, such as the proximity of local shops, schools, leisure facilities and other amenities. It may be worthwhile to establish how good the local schools are and the level of crime in the area. This will help to establish the marketability of the house or houses that are proposed to be built on the land.

Other useful information will be the sourcing of local building materials suppliers, plant hire companies and local tradesmen. In addition it can also be determined where excess soil from construction operations could be deposited and where waste from the site could be tipped.

The physical or site study

On arriving at the site it will be possible to establish the quality of access to the site. Is it possible to use the existing access or will it need to be widened to get materials and equipment onto the site? It will also be useful to note the orientation of the land (i.e. the direction in which it faces), the prevailing climate in the area, the predominant direction of the wind and typical wind speeds, how exposed the site is, the level of rainfall, the amount of frosts in the area and the amount of sunshine, and average daytime and night-time temperatures. Much of this information can be obtained from a desk study before arrival on site. It may be possible to find out information about the site from local inhabitants such as what had been built on it previously, did the land drain well after prolonged periods of rainfall and was the land used for landfill at any time? If there are any structures on the land it is useful to investigate these for signs of any structural damage that may have been caused by settlement or subsidence. Neighbouring properties can also be observed for any signs of damage. If the land has previously been developed it is important to discover whether any services, tanks or old foundations remain on the land.

Self-assessment question 3.3

Why should it be important to discover whether any services or old foundations remain on the land?

Following this it is important to walk over the site and note the general topography (i.e. the shape of the land) and changes in level of the land. In addition there can be tell-tale signs on the surface of the land that can indicate the type of ground beneath, such as the presence of marsh grass, indicating a high water table; the stickiness of the ground, indicating clay subsoil; the presence of rubble, indicating a possible previous use as landfill. The type of vegetation on the land should also be noted. To what extent are there trees and bushes on the land, and their sizes? Some trees may have a preservation order imposed on them and this should have been previously checked out when undertaking the desk study.

Having undertaken the walk over survey it will then be necessary to investigate more about the type of ground that the land is made up of.

The type of ground

The type of ground that the building is being constructed upon can be basically classified into two categories:

- Rocks – hard, rigid and strongly cemented deposits with a high loadbearing capacity, but difficult to excavate.

- Soils – soft, loose and uncemented deposits that are easier to excavate than rocks, but generally have a lower loadbearing capacity.

  Rocks are generally of three distinct types:

- Igneous rocks – for example granite and basalt. These are formed by the solidification of molten material. They have the highest loadbearing capacity, which is two to three times that of sedimentary rocks and 25–50 times that of soils.
- Metamorphic rocks – for example slates and marbles. These are formed from consolidated deposits altered by heat or pressure. They are hard, but subject to faults that can allow movements, which can affect the stability of any building constructed above them.
- Sedimentary rocks – for example sandstone and limestone. These are formed by deposits cemented in layers or strata. The quality of the rock is dependent upon the quality of the cementing material, the angle of stratification and the behaviour of the rock in wet conditions. Large cavities can be caused by the passage of water through soft rock.

The loadbearing capacity of rocks is reduced if they are not in sound condition due to:

- weathering
- shattering through the effects of earth movements
- steeply dipping bed joints
- soft clay deposits occurring in bed joints.

Soils consist of individual mineral particles that surround spaces or voids. These voids may be filled with air and/or water. The mineral particles can support load and resist shear stresses. The water in the voids can support some load, but cannot resist shear stress and may be squeezed out of the voids under heavy pressure. Air in the voids cannot support load and will be squeezed out of the voids under relatively light pressure.

Table 3.1 The loadbearing capacities of rocks

Type of rock	Safe bearing capacity in kN/m^2
Strong unweathered igneous & gneissic rocks	10000
Unweathered limestones & sandstones	4000
Schists and slates	3000
Weak shales, mudstones & sandstones	2000
Weak weathered shales & mudstones	600–1000
Weathered chalk & limestone	600

Adapted from *BS 8004:1986 Code of practice for foundations*. Reproduced by permission of BSI.

Soils may be classified into many different types, but a general classification by size and nature of particles, to assess the density and structural properties of the soil, is generally used. There are four broad types of soil:

- Non-cohesive – for example, sands or gravels
- Cohesive – for example, clays and silts
- Organic – for example, peat
- Synthetic or man-made – soils that have been imported from a different location or materials used for landfill, such as domestic refuse or demolition rubble.

Many soils may be a mixture of two or more types.

BS 5930:1999 + Amendment 2:2010 Code of practice for site investigation specifies coarse grained soils as having 65 per cent or more of particles having a size larger than 60µm (microns, that is one thousandth of a millimetre), and fine grained as having 35 per cent or more of particles having a size smaller than 60µm.

Non-cohesive soils have large, coarse grained particles that range from 2–60mm in diameter for gravels and 60µm–2mm in diameter for sands. The particles are also irregular in shape and, because of their size, the voids between the particles tend to be large, leading to well-drained soils. This property will largely depend on the range of sizes of particles within the soil (referred to as *grading*). In a well-graded soil smaller particles will fit into the voids between larger particles, making it easier to compact the soil, and the density of the soil will be increased.

Self-assessment question 3.4

Why does a well-graded soil generally have a better loadbearing capacity than a soil that is poorly graded?

Because the particles are irregular in shape and coarse in texture, they tend to exhibit high frictional resistance when compressed under load and this feature contributes significantly to their loadbearing capacity. However, the particles do not bond together well if they are not consolidated, thus they will have low loadbearing capacity if they are in a saturated state, caused by a high ground water level separating the particles and reducing their frictional resistance (consider quicksand). The sides of excavations in these types of soils will also need a great deal of support.

Self-assessment question 3.5

Why should frictional resistance contribute significantly to the loadbearing capacity of non-cohesive soils?

Table 3.2 The loadbearing capacities of non-cohesive soils

Type of non-cohesive soil	Typical loadbearing capacity kN/m² dry
Dense gravel or dense sand & gravel	≥600
Medium dense gravel or medium dense sand & gravel	200–600
Loose gravel or loose sand & gravel	≥200
Dense sand	≥300
Medium dense sand	100–300
Loose sand	≤100

Adapted from *BS 8004:1986 Code of practice for foundations*. Reproduced by permission of BSI.

Cohesive soils, on the other hand, tend to have smaller, smooth grained particles that range from 60μm to microscopic in size. The particles tend to be regular in shape, resembling flat, plate-like structures. They also possess a small negative electrical charge. This charge attracts the positive charge of water molecules (water molecules are dipolar, that is they have both a positive and negative electrical charge) and this binds the particles of the soil together.

The water present in these soils gives them their plastic characteristics (that is they are easily moulded under load but will not return to their former shape once the load has been removed) and makes them sticky when wet and crumbly when dry. Therefore the seasonal changes in the water content of these soils will cause them to swell in the winter months when rainfall is high and to shrink during the summer months when rainfall is low. Shrinkable soils having a Plasticity Index between 10–20 per cent are considered to have low shrinkage potential, those having a Plasticity Index of 20–40 per cent are considered to have medium shrinkage potential and those having a Plasticity Index of above 40 per cent are considered to have high shrinkage potential.

Self-assessment question 3.6

Why should the plasticity of a cohesive soil be of concern?

They will also be subjected to frost heave when the water between the particles that are close to the surface of the ground freeze during cold weather. In addition to this, they will also be affected by the roots from nearby trees drawing water from the soil during periods of dry weather. Shrinkable clay soils, especially those located in the area around the south-east of England, can suffer from these problems. These aspects are considered further in Chapter 5 (foundations).

Self-assessment question 3.7

Why should cohesive soils affected by frost during cold weather suffer from frost heave?

Cohesive soils rely heavily on the bond between the particles and the molecules of water binding them together. This, together with their small void spaces due to their small particle size, helps to explain why these soils do not drain as easily as non-cohesive soils. The loadbearing capacity of cohesive soils is also related to their moisture content. Too little water in the soil causes the soil to crack and crumble, too much water causes the soil to become too slippery with high plasticity, and shear failure may occur before the soil reaches its normal loadbearing capacity. However, where the moisture content of a cohesive soil is close to its optimum level, the sides of an excavation may only require minimal support. Under the compression exerted by a building load, water will be gradually squeezed out from the voids in these soils leading to long-term settlement. This aspect will be considered further in Chapter 5 (foundations).

Self-assessment question 3.8

Why should cohesive soils, where the moisture content is close to the optimum level, require minimal support in an excavation?

Table 3.3 The loadbearing capacities of cohesive soils

Type of soil	Typical loadbearing capacity in kN/m^2
Very stiff boulder clays & hard clays	300–600
Stiff clays	150–300
Firm clays	75–150
Soft clays & silts	≤75
Very soft clays & silts	Not applicable

Adapted from *BS 8004:1986 Code of practice for foundations*. Reproduced by permission of BSI.

Organic, fibrous soils contain a high amount of organic matter and their volume varies considerably with the moisture content. These soils generally have a high amount of voids and are therefore highly compressible so that they settle readily when loaded and have a relatively poor loadbearing capacity. Peats are an obvious example of this type of soil and apart from their high compressibility they are also highly acidic, which can produce an aggressive environment for foundations and services that are in contact with these soils. They can also give off methane gas which is highly flammable, can have an unpleasant smell and is often harmful to occupants of the site.

Self-assessment question 3.9

Why should highly acidic soils produce the possibility of an aggressive environment for foundations?

Topsoil will generally make up the top 150–300mm of soil on a site. This soil contains a large amount of decaying plant matter and is also highly compressible. Although it is good for growing plants in, it is not suitable for bearing loads. It must therefore be removed and stored on site for later landscaping purposes before construction work can commence. This will be discussed in more detail in Chapter 4 (excavations).

Synthetic or man-made soils are often encountered on sites previously used for landfill purposes. As the availability of land that may be used for construction in this country decreases, then more use has to be made of land that has been restored for construction use following its use for refuse disposal. One of the important factors to be considered for these soils is what they consist of and how well they have been compacted during the landfill operations and the strength of the underlying soil. Some materials used in landfill are organic in nature and may well emit methane gas as they decay. This is highly flammable and must be vented adequately before construction work can commence. Other materials may corrode over time, leaving voids in the fill.

Self-assessment question 3.10

Why are voids from decayed or corroded materials in the fill a problem?

On sites of varying topography, where there are wide ranges of differing levels, subsoil may be imported from other areas to fill up the dips and hollows. These imported soils may differ considerably in structure from the naturally occurring soils on the site and their structural behaviour is likely to be strongly influenced by the amount they are compacted in place during the filling operations as well as the strength of the underlying soils.

Contaminated ground

As mentioned earlier, it is important to establish the previous usage of the land as part of the site investigation. If the land has been used for industrial purposes then the ground may contain toxins that are likely to be harmful to people using the land, either through emissions into the buildings constructed on the land or through the pollution of any vegetation grown on the land. Even when the land has previously been used for agriculture or forestry, the land can have become contaminated by pesticides, fertilisers, fuel and oil, or decaying biological matter. Natural contaminants may already exist on the land due to the underlying geology. These include heavy metals and gases such as methane and carbon dioxide, originating in coal mining areas and where organic soils such as peat are found. Natural contaminants also include the radioactive gas radon. This is a colourless and odourless gas formed by the decay of radioactive uranium and radium deposits. It can percolate through the soil and into buildings. Some areas of the country are more susceptible to radon than others. Where buildings have previously been erected on the land it is important to check whether subterranean structures such as foundations, tanks and service pipes and cables are still present. These will need to be removed before building work commences. Ground water present on the site can adversely affect the stability and properties of the ground and if contaminants are present they could be transported into the building or the services by the ground water movement. Where the water table can rise to within 250mm of the ground floor of the building, the land should be drained to a suitable water course by the use of porous pipes.

The land should be checked for contamination through the site investigation process previously discussed. Where contamination is found, eradication strategies will depend on the type and quantity of the contaminant. The contaminant may be treated by the use of physical, chemical or biological processes to reduce its toxicity or harmful properties. The contaminant may also be isolated beneath protective layers or the installation of barriers to prevent migration within ground water. The contaminant may also be removed by excavation and disposal onto a suitably licensed landfill site. It may also be possible to re-design the layout of the site so that any contaminated ground can be sealed beneath a hard landscaped area, such as paving. If soil has to be imported onto the site to make up levels, its freedom from contaminants needs to be verified. Where gas contaminants are detected, a gas resistant barrier will need to be placed across the entire footprint of the buildings to be erected and a ventilation layer added from which the gases may be dispersed to the atmosphere. The positioning of this membrane may affect the type and the positioning of insulation in the ground floor construction.

Soil investigations

Information from the desk study and the site walk over should provide useful information concerning the nature of the ground on the proposed site. Where the land has previously been built on it may well be possible to ascertain the nature and engineering characteristics of the ground without further costly investigation. Where this is not possible however, further information on the type of ground and its engineering characteristics will need to be obtained from tests on the ground itself.

These tests can be either undertaken in a laboratory or on site. It is generally accepted that laboratory tests can be undertaken with precision equipment in favourable surroundings and are therefore likely to be able to provide more precise results than tests undertaken on sites. However, laboratory tests can only be undertaken on samples of the ground obtained from the site, whereas site tests can be undertaken on the ground as it occurs 'in situ'. In practice, information from both sources is useful in order to build up a picture of the ground on the site being investigated.

In order to undertake either site or laboratory-based tests, access to soil samples must be obtained. The three most commonly used methods are:

- the trial pit
- the auger
- the core sampler.

The trial pit

This is a pit normally excavated by machine and of sufficient depth to provide useful information on the type of soil likely to be encountered in the foundation excavations and onto which the foundations will be bearing their loads. The trial pits should be large enough to allow a physical examination of the soil within and also allow site tests of the soil to be undertaken. It is possible for samples of the soil excavated from the trial pit to be sent to the laboratory for further analysis.

The auger

This is a borehole excavated by a helically-shaped drill, which is normally mounted on a lorry or tractor. The borehole can be taken to much greater depths than the trial pit, but it is not possible to physically examine the soil in situ. Samples from the borehole are generally sent to the laboratory for analysis. The disadvantage of laboratory samples that are obtained from trial pits and boreholes is that they are disturbed (i.e. the soil from different levels may be mixed up in the sample sent to the laboratory for testing). However, one advantage of the borehole is that because it can be taken to much greater depths than the trial pit it is possible to establish the level of the natural water table on the site.

Self-assessment question 3.11

Why is a disturbed soil sample less useful than an undisturbed soil sample?

The core sampler

This is a cylindrical tube with a cutting edge that can be driven into the ground to obtain core samples of the soil. These core samples can then be removed from the tube and sent to the laboratory for investigation.

The samples are undisturbed and can therefore give the best opportunity for laboratory tests to be undertaken on the soil as it would appear in situ.

The number and location of trial pits, auger boreholes or core samples will be dependent on the nature of the site, the variability of the ground and the size and location of the proposed buildings.

Soil testing

As mentioned previously there are two methods of testing soils:

- site testing
- laboratory testing.

Site testing

The Standard Penetration Test measures the resistance of the soil under loading. A 35mm *split spoon sampler* is driven into the soil at the bottom of a borehole. The sampler is initially driven a distance of 150mm into the soil by means of a standard weight of 63.5kg falling through a set distance of 750mm. The sampler is then driven a further 300mm into the soil and the number of 'standard' blows needed to achieve this is recorded. From this information the relative density of the soil can be established.

The Vane Test measures the shear strength of soft cohesive soils in situ. The apparatus comprises a 'vane' of two 75 x 38mm plates arranged at right angles to each other in a cruciform pattern. The plates are attached to a rod that has a *torque* measuring gauge at its head. The vane is pushed into the soft cohesive soil and rotated by hand at a constant rate until the soil within the vane fails by shearing. The torque necessary for shear failure to occur is measured from the gauge.

The Plate Bearing Test involves the loading of a 600 x 600mm steel plate at the bottom of a trial pit. The load can either be applied by *kentledge,* involving a large number of heavy concrete blocks, or alternatively a hydraulic jack can transmit the load by being jacked against the underside of a heavily loaded beam that is positioned at ground level. The load is applied in increments of 20 per cent of the design load at 24-hourly intervals and the settlement of the soil is measured every six hours and plotted on a time graph. Failure under load is assumed when the settlement reaches a depth equal to 10 per cent of the breadth of the loading plate (i.e. 60mm). The safe bearing load can then be assumed to be one-third of the load that has caused failure under the test.

Laboratory testing

Soil samples are identified and classified when they first arrive at the laboratory for testing. A visual examination determines the colour, texture and consistency of the samples. The moisture content of the samples is then measured. The greater the moisture content of the soil, the higher will be its compressibility; thus it is important to ascertain the liquid and plastic limits of cohesive soil samples. The liquid limit determines the moisture content needed to cause the soil to flow under a given number of vibrations. The plastic limit is determined by rolling out a 4mm-diameter thread of the soil and noting the moisture content which will allow the thread to be rolled out still further until it breaks due to drying. The Plasticity Index of the specimen can be determined by subtracting the moisture content of the plastic limit from that of the liquid limit. The higher the Plasticity Index of the sample the greater will be the plasticity of the soil. Finally, by sifting dried samples of soil through a series of various sized sieves it is possible to determine the particle size distribution or grading of the sample.

The shear strength of a cohesive soil can be measured by the Triaxial Compression Test. A cylindrical specimen 75mm long and 38mm in diameter is taken from an undisturbed sample and placed in a plastic cylinder that is then filled with water. The specimen is then subjected to lateral hydraulic pressure as well as a vertical load and the force needed to shear the specimen is measured. The test is repeated on two more specimens taken from the same undisturbed sample and each is subjected to a higher hydraulic pressure before the vertical load is applied. The results are plotted as *Mohr's circles* and the tangent that touches the circles obtained from all three specimens is drawn from the vertical axis of the graph. The angle that this tangent subtends from the horizontal determines the increase of shear strength with load.

Self-assessment question 3.12

What type of soil will not have an increase of shear strength under load?

For non-cohesive soils the shear strength can be measured by the Shear Box test. A sample of soil is placed into the box of the apparatus and subjected to a standard load whereby a horizontal force is applied to the lower half of the box until the sample fails under shear.

The magnitude and rate of consolidation of a soil sample can be measured by an odeometer. A cylindrical specimen of soil 75mm in diameter and 18mm thick is placed in a metal ring and capped with porous discs. It is then placed in a tray filled with water and subjected to a vertical load. The load is increased every 24 hours and a time/settlement curve is plotted.

Finally a chemical analysis of the soil and any ground water collected in the samples is undertaken. The main information that needs to be ascertained is the sulphate salt content and pH value of the soil and ground water, together with the existence of any contaminants.

Visit the companion website to test your understanding of Chapter 3 with a multiple choice questionnaire.

CHAPTER

Excavations

Introduction

Before foundations can be prepared for the construction of a building, excavations will need to be dug to accommodate them. The size of these excavations will be dependent on three factors:

- the type of ground
- the type of foundation
- the type of building.

Before excavation can commence, the shape of the building needs to be established on the site and the position and depth of excavations needs to be ascertained. A short section on setting out is therefore included. This topic is covered in more detail in books on site surveying.

In the section on excavation itself, consideration is given to the main types of excavation undertaken on construction sites before considering the types of excavation plant that are generally used. Most types of excavation plant are more efficient at certain types of excavation operation than others and this aspect is considered in the review of each type of plant. The range of machinery available for excavation operations is immense and some of these machines may not be particularly suited to residential development sites. The review of machinery acknowledges this and so only those machines that are likely to be regularly encountered on residential development sites are considered.

Setting out

Setting out involves transferring details of the positioning of the building from a drawing onto the site. It requires a number of setting out lines and pegs. The pegs are positioned away from construction activities in order to ensure that they remain undisturbed.

Buildings are positioned in relation to roads or other boundary lines. The *Building Line* is often established by the local planning authority, in front of which no building work can normally be undertaken; this therefore signifies the position of the front face of the building. This can be set out by taking offsets (measurements at 90^0) from the road. Corners of the building are denoted by pegs, which are set out by use of surveying instruments, such as a theodolite, and tape. A favourite technique for establishing a right-angled corner with the use of a tape is to use the 3–4–5 rule. This uses Pythagoras's theorem that a triangle having two sides of three and four units in length respectively, and with a hypotenuse of five units in length, must have a right angle that is opposite to the hypotenuse.

More complex-shaped buildings, with little or no reference details, may need to be set out by a site engineer. With the use of modern total stations and a construction drawing with adequate information so as to extract co-ordinates, the engineer will be able to establish new base control points closer to the proposed building.

This will allow setting out to be carried out by less complex methods at a later stage. Alternatively the site engineer may set out proposed trench centre lines ready for excavation using bearing and distances from the newly positioned control points. One advantage of this method of construction is that perimeter site vegetation, sometimes crucial to planning approval, need not be affected in order to facilitate accurate positioning of the building.

Setting out control pegs are often destroyed during the early stages of the construction process through the use of earth-moving vehicles. Often a centre line of trench pegs is destroyed when the foundations are being excavated. It is therefore important that the control points which have been established on site are adequately protected, as the re-establishment of these positions can be timely and expensive with the re-employment of a site engineer (see Figure 4.1).

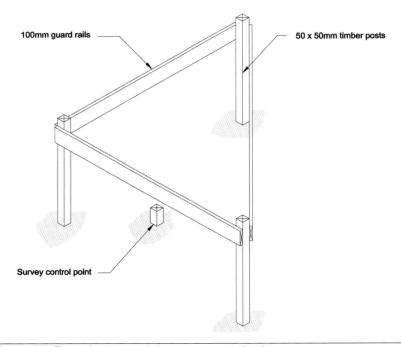

100mm guard rails

50 x 50mm timber posts

Survey control point

Figure 4.1 Protection to an existing survey control point

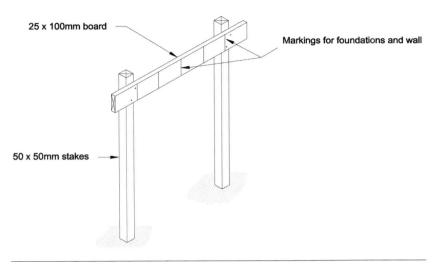

25 x 100mm board

Markings for foundations and wall

50 x 50mm stakes

Figure 4.2 Profile board

If it is inevitable that a peg's position is to be destroyed, a configuration of reference pegs can be adopted. Four pegs can be positioned so that the intersection of a string line drawn between each pair of pegs will coincide directly above the centre of the control peg. Distances should be taken between the reference pegs and also between the control peg and the reference pegs to check against movement.

All information, graphical or numerical, relating to the reference points should be recorded, so as to aid in the re-establishment of the control points and to check the accuracy of the works.

Having set out the shape of the building *profile boards* will be erected. These denote the positions of trenches and have the centre lines of walls and foundations marked upon them. They consist of a horizontal timber rail attached to two vertical timber posts, which are driven into the ground (see Figure 4.2).

Not only must the position of trenches be located, but the depths of the excavations must also be ascertained. The profile boards can therefore be set at a suitable height above the ground at each end of the trench and a *traveller,* comprising a vertical standard with a horizontal cross rail set at the correct height (i.e. the proposed depth of the trench plus the height of the profile board above the ground level), is then placed on the bottom of the trench and sighted between the two profile boards. The correct depth of trench is achieved when the two profile boards and the cross rail of the traveller coincide on the sight line (see Figure 4.3).

Excavations

There are three main types of excavation on building sites:

- general earth-moving
- trench excavation
- bulk excavation.

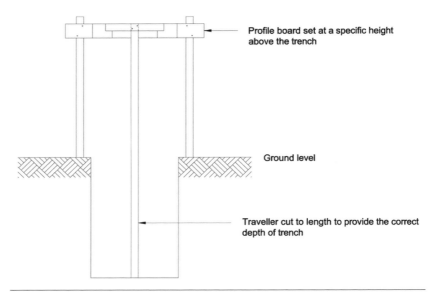

Profile board set at a specific height above the trench

Ground level

Traveller cut to length to provide the correct depth of trench

Figure 4.3 Establishing the correct depth of the trench using a traveller

Most excavation undertaken on building sites nowadays is carried out by machine. Hand excavation is extremely slow and expensive and particular care needs to be exercised with the safety of construction personnel within excavations. However, there may be certain situations where hand excavation is more suitable than excavation by machine. This may be where access to the excavation is particularly difficult for a machine (this is becoming less of a problem as smaller micro and mini excavators are being developed), or where particular obstructions such as tightly packed service pipes and cables or archaeological remains are likely to be encountered. There may also be circumstances where the amount of excavation is so small that it would be uneconomical to use a machine and in these situations hand excavation may be considered preferable. These situations are likely to be the exception rather than the rule.

Each type of excavation uses its own type of machinery.

General earth-moving

This would involve:

- general site clearance operations such as removal of vegetation and rubble
- stripping of topsoil and mounding it for later use in landscaping operations
- reduced level dig to reach the formation level for the start of construction
- cut and fill operations where the topography of the site varies.

These operations require a machine that can move soil, vegetation and rubble from a relatively large area and either push the material into a

Figure 4.4 Bulldozer

mound in the corner of the site or be able to load the material into transport that will remove it from the site. The depth of excavation is relatively shallow and not likely to exceed 300mm. Ideal machines for these operations will have a curved front blade for stripping and pushing material in front of the machine (a bulldozer, see Figure 4.4) or have a front bucket that is able to strip soil, grub up vegetation, mound material or load material into the back of a lorry for removal from the site (a tractor shovel, see Figure 4.5). Where trees and shrubs need to be removed from the site it is important to ensure that their roots are grubbed up effectively, so that they do not interfere with later construction, particularly foundations.

These machines rely on their 'tractive efficiency' to put as much of their power as possible into the stripping and pushing operations that they perform. This relies on the machines being able to grip the ground that they are travelling over as firmly as possible without slipping. For this reason most of these machines will have 'caterpillar' tracks rather than wheels with pneumatic tyres. The 'caterpillar' tracks also provide enhanced manoeuvrability to the machine, but they travel at much slower speeds than vehicles with pneumatic tyres and they are unable to travel on public roads, due to the damage they would do to the road surface; so they need to be transported to and from site on a low loader.

Figure 4.5 Tractor shovel

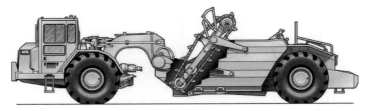

Figure 4.6 Scraper

Self-assessment question 4.1

Why can caterpillar tracks provide enhanced manoeuvrability to the machine?

On very large sites where the topography is such that there may be considerable differences in the level of the ground across the site, it may be necessary to undertake 'cut and fill' operations. This involves removing the surplus soil from the high points and using this to fill the low points in an attempt to produce a level site. The scraper is an ideal machine for this operation (see Figure 4.6). It comprises a large 'bowl' that has a front cutting edge that can be lowered to the appropriate level of cut. The 'bowl' is towed around the site by a tractor unit and works on a similar principle to a pencil sharpener: that is it takes a shaving from the ground as the cutting edge passes over it, which is then fed into the bowl. The bowl may then have its contents emptied by raising its front bulkhead and spreading the soil as the machine moves forward. These machines need a very large turning circle and are therefore not particularly useful on compact sites.

Trench excavation

Trench excavation involves digging in long lines. The depth of the excavation is generally greater than its width. On housing sites, trenches are dug for the installation of linear foundations or service pipes and cables. The depth would generally not exceed 2m and the width would generally not exceed 600mm for residential development projects. Suitable machines are the backhoe, a tractor with a rear bucket having a mechanism similar to the human arm, with joints at the shoulder, elbow and wrist (see Figure 4.7). The excavator arm has detachable buckets for different widths of excavation. The machine has a front bucket for loading loose materials. The tractor has wheels with pneumatic tyres, which allows it to travel independently between sites and to move relatively swiftly on site. However, these wheels are not particularly suitable in very muddy conditions. Another disadvantage with this machine is that the depth of dig is confined to a maximum of approximately 4.5m and the backhoe can only operate in an 180⁰ arc behind the machine. Neither of these restrictions limits the use of this machine on most housing sites.

Figure 4.7 Backhoe

Self-assessment question 4.2

Why are caterpillar tracks better than wheels for vehicles on muddy sites?

Alternatively, a backactor may be used for trench excavation. This machine has an excavation arm similar to that of the backhoe, but the arm is attached to an operator's cab that is mounted on a chassis that has 'caterpillar' tracks (see Figure 4.8). The cab and excavator arm are therefore able to operate through a 360^0 arc around the machine and the machine is able to work in poor ground conditions. The excavator arm of the backactor is also capable of digging to greater depths (up to 9.5m) and can accommodate much wider bucket sizes than that of the backhoe. Both the backhoe and backactor are capable of excavating and loading.

For much smaller excavations or for trench excavations in locations that are inaccessible to larger machines, a variety of mini-excavators and micro-excavators have now been developed (see Figures 4.9 and 4.10).

For excavating and backfilling shallow and narrow trenches for the installation of service pipes and cables, the trencher has been developed. This machine operates on the conveyor belt principle, having a series of small excavating buckets attached to continuous loop chains that are supported by a boom. The excavated spoil is deposited alongside the trench by plough-shaped deflectors mounted on the machine. These

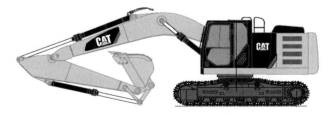

Figure 4.8 Backactor

Figure 4.9 Mini-excavator

Figure 4.10 Micro-excavator

machines can be adapted to excavate the trench, lay the service pipe or cable and backfill the spoil in one continuous operation with considerable speed and accuracy.

Bulk excavation

This involves deeper and wider excavations than those required with trench excavation. On a residential development project this form of excavation would be used for the construction of single storey basements.

If the excavation of the basement is not too large, a backhoe may be able to undertake the excavation, otherwise a backactor would be used. Another option is to use a face shovel. This machine is similar to a backactor but the excavator arm works in the opposite direction, away from the cab (see Figure 4.11). It operates by digging into the cliff face ahead of the machine and as it cannot excavate below the level of the machine it must be located at formation level to dig out a basement. In order to reach the formation level and also to allow the lorries taking the excavated soil from the site to reach the excavator, the excavation would need to incorporate a ramp that has a relatively shallow incline. This requires significant space and would normally only be available on large basement excavations that are unlikely to be found on residential development projects.

Figure 4.11 Face shovel

Self-assessment question 4.3

Why should the ramp require a shallow incline?

Earthwork supports

Earthwork supports are used to retain the sides of the excavation to prevent them from collapsing during the period that the excavation remains open and provide:

* safe conditions for any construction personnel to work within the excavations
* prevent damage to adjacent structures that could be caused by the excavations
* enable work to proceed within the excavations without interruption.

In providing support to an excavation a number of factors need to be taken into consideration:

* the nature of the soil – generally non-cohesive soils require more support than cohesive soils do.
* the depth of the excavation – shallow excavations need less support than deep excavations.
* the width of the excavation – wide excavations will need to be supported in a different way to that of narrow excavations.
* the type of work to be carried out – operations carried out within the excavation will require working space. The amount required will be dependent on the operations involved.
* the moisture content of the ground – soils will require different amounts of support as changes in their moisture content occur.
* the length of time the excavation will be left open – cohesive soils, in particular, may dry out and start to crumble if the excavation is left open for long periods in dry weather.
* the method of excavation – hand excavation will require more support than machine excavation.

- the support system used – different methods of excavation support can be installed before, during or after the excavation.
- the removal of the support system – different support systems can be removed either before or after backfilling operations.
- moving materials into excavations – the working space will need to consider the materials being moved into and out of the excavation as well as the operations being carried out within the excavation.
- the proximity of existing buildings near to the excavation – applied loads from foundations can stress soils at the sides of excavations.
- the use of land adjacent to the excavation for stacking materials – over-loading of the ground by stacking materials close to the excavation can cause stress on soils at the side of the excavation. Similarly large vehicles should be prevented from driving too close to the excavation.
- vibration of soils from construction operations or vehicles using adjacent roads – excessive vibrations can cause soils to move, making the sides of an excavation potentially less stable.

It is possible to retain the sides of an excavation without support by sloping the sides of the excavation to the *angle of repose* of the soil. This is the natural angle at which the soil will remain stable without additional support. When soil is tipped into a mound it settles to its natural angle of repose. This angle will alter according to the type of soil and its moisture content. Because the angle of repose can be rather shallow, this method of excavation support does take up a large amount of space on site and is therefore not frequently adopted. However, this technique can be used on bulk excavations.

The temporary support generally used in excavations is called *planking and strutting* and comprises:

- vertical boards called *poling boards* to the faces of the excavation
- horizontal boards called *sheeting boards* to the faces of the excavation
- horizontal bearers called *walings* spanning across the poling boards to support them, down the length of the excavation
- horizontal *struts* spanning across the excavation between opposing poling boards or walings to resist the pressure from the soil
- tapered *wedges* used to tighten the struts to the poling boards or walings.

There are four main types of planking and strutting:

- full or close sheeting
- half sheeting
- quarter sheeting
- one-eighth or open sheeting.

Full or close sheeting would generally be used in loose wet soils, such as very soft clays or silts and loose uniform sands or gravels and comprises vertical poling boards driven into the ground to a depth of

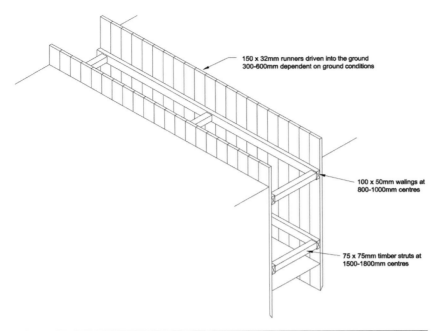

150 x 32mm runners driven into the ground
300-600mm dependent on ground conditions

100 x 50mm walings at
800-1000mm centres

75 x 75mm timber struts at
1500-1800mm centres

Figure 4.12 Full or close sheeting

300–600mm, dependent on conditions (called *runners*) and supported by walings at 1m centres with struts wedged between them at 1.8m centres horizontally and 1m centres vertically (see Figure 4.12).

Self-assessment question 4.4

Why should the runners need to be driven into the ground at the base of the trench?

Half sheeting would generally be used in firm, loose, or loose and wet soils, such as soft clays and silts and compact or loose well-graded sands or gravels, and comprises poling boards fixed alternately along the side of the trench (i.e. the space between boards is equivalent to the width of one board) supported by walings at 1m centres with struts wedged between them at 1.8m centres horizontally and 1m centres vertically (see Figure 4.13).

Quarter sheeting would generally be used in firm soils, such as firm clays and comprises poling boards with spaces equivalent to two board widths between them, supported by walings at 1m centres with struts wedged between them at 2–3m centres horizontally and 1m centres vertically (see Figure 4.14).

One-eighth or open sheeting would generally be used in hard soils that require little additional support, such as stiff clays. It comprises poling boards with spaces between them equivalent to the width of four boards with struts wedged between them (see Figure 4.15).

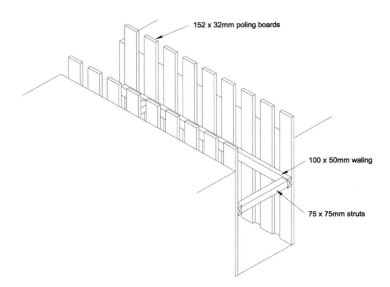

Figure 4.13 Half sheeting

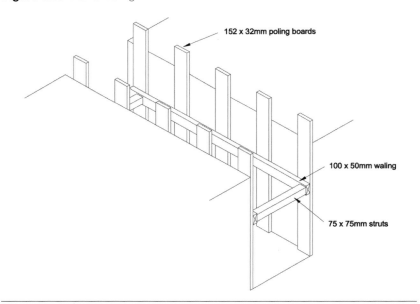

Figure 4.14 Quarter sheeting

Self-assessment question 4.5

Assuming the width of a normal poling board is 300mm what would be the spacing of poling boards in:

a) half sheeting?

b) quarter sheeting?

c) one-eighth or open sheeting?

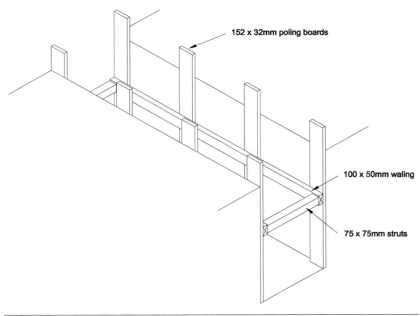

Figure 4.15 One-eighth or open sheeting

Traditionally timber has been used for trench support components, but increasingly steel trench sheets are replacing timber poling boards and steel telescopic struts are replacing timber struts and wedges (see Figure 4.16).

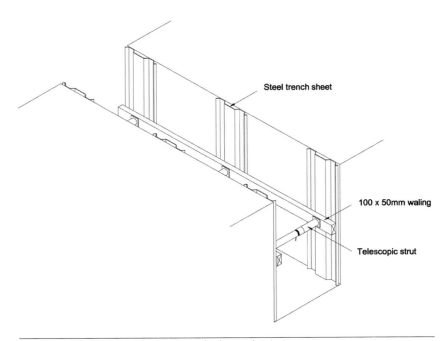

Figure 4.16 Steel trench sheets and telescopic struts

In addition, proprietary hydraulically operated frames, boxes and plate lining systems are increasingly being used for excavation supports (see Figure 4.17). They have the advantages of being quick to install and remove and can be installed from above the excavation without personnel having to enter an unsupported excavation to install the support system.

Figure 4.17 Proprietary hydraulically operated frame

Visit the companion website to test your understanding of Chapter 4 with a multiple choice questionnaire, and to see the following features illustrated with full colour photos:

- **4.1** Gravel lined drainage trench
- **4.2** Mini-excavator
- **4.3** Plastic cellular boxes forming storm water attenuation tank

CHAPTER 5

Foundations

Introduction

In this chapter it is intended to consider foundations suitable for a lightly loaded structure such as a house. In so doing the interaction between the foundations and the loadbearing walls as well as the foundations and the soil beneath the building will be discussed. In particular the effect of settlement and the effect of tree roots on foundations will be considered.

The primary function of a foundation is to transmit all dead, live and superimposed loads from the building to the surrounding ground, such that excessive settlement of the structure and failure of the soil to support the load is avoided.

Foundations should be constructed at such a depth as to avoid damage by swelling, shrinkage or freezing of the subsoil and be capable of resisting attack by sulphate salts that may be present in the soil. The effect of sulphate salts on concrete foundations will be considered further in Chapter 6 (concrete).

Choice of foundation

The choice of foundation for a particular house depends mainly on three factors:

- the total loads of the building
- the nature and bearing capacity of the subsoil
- the amount of settlement produced by the loading.

The total loads of the building

As discussed in Chapter 1 (introduction), there are three main types of loading on a building; dead load, live load and superimposed load. The total load on the foundations is summated from these individual loads. It is assumed that the loading imposed by a house on its foundations is uniform all around its perimeter. This is not strictly correct, since live loadings from floors and superimposed loadings from roofs will only be borne by the walls that these elements bear upon. However, to reduce the complexity of calculating individual loadings for each section of the external walls of a house, it is easier to take the worst case scenario (i.e. the wall experiencing the greatest sum of loads) and assume that all other walls are experiencing the same load. If the load is assumed to be uniform, then it is not necessary to add up all the loads for the entire perimeter of the house. If a sample 1m of the external wall is taken, then the total load on that section of wall can be determined and the foundation needed to support that section of wall can be calculated. This design can then be applied to the rest of the foundations for the house in question.

Self-assessment question 5.1

Why should a 1m sample of wall be taken to calculate the size of the foundation to support it?

The nature and bearing capacity of the subsoil

As discussed in Chapter 3 (site investigation and the type of ground), the nature and bearing capacity of the subsoil (i.e. the soil beneath the topsoil) varies with the type of soil, its degree of compressibility and the amount of moisture in the soil. In addition, cohesive soils, particularly clays, can be subject to seasonal movement up to a depth of 1m. These soils may be subjected to shrinkage or contraction in summer months and swelling or expansion during winter months.

Self-assessment question 5.2

Why should seasonal movement be confined to 1m below ground level?

The amount of settlement produced by the loading

Soil is compressible to varying degrees. As a load is applied to a foundation then the soil beneath the foundation will be compressed, the water and air in the voids between the particles will be squeezed out and the foundation will settle. This process of consolidation will continue until the forces between the particles are equal to the applied load. The speed of this consolidation or settlement is determined by the speed of the migration of the water and air from between the soil particles. Foundations built on sands therefore settle relatively rapidly, whilst the settlement of foundations built on clay soils is much slower and can last for a number of years.

Soils that are close to the surface are likely to be more compressible than those at greater depths, simply because deeper soils have been compressed by the weight of the soil above them. If the applied load on a clay soil is reduced by excavation, water tends to move to the unloaded areas and swelling of the soil occurs.

Peats and other soils containing a lot of organic matter shrink and swell easily as their water content changes. They are very compressible and settle readily even under light loading. Made up ground behaves in a similar manner unless the material is well graded, carefully placed and properly compacted in thin layers. Shallow foundations should not be used on sites consisting of made up ground.

Slight settlement should not cause problems to the structure of the building. Excessive settlement may cause shear failure of the soil.

Settlement must also be uniform throughout the building otherwise damage may result from differential settlement. The amount of differential movement between parts of a building must be kept within acceptable limits.

Self-assessment question 5.3

Why is excessive differential settlement harmful to a building?

Types of foundation

Foundations may be divided into two broad categories:

- shallow foundations
- deep foundations.

Shallow foundations

These may be strip, pad or raft foundations. They are described as shallow foundations because they transfer the load of the building to the subsoil at a level close to the surface. They are nearly always the cheapest to install and are generally used where sufficient depth of strong subsoil exists near to the surface of the ground. The foundation needs to be designed so that the soil is not over-stressed and so that the pressure on the subsoil beneath is equal at all points in order to avoid unequal settlement.

If a diagram were to be drawn linking the points of equal pressure within the soil beneath a foundation it would resemble a circular 'bulb' emanating from the base of the foundation and extending into the ground for a considerable depth. This distribution of pressure into the ground may be likened to sound waves emanating from a loudspeaker. The intensity of the sound decreases the further away from the source the sound waves travel, and it is the same with pressure being applied to the soil due to the building load. When considering the pressure on the soil from the foundation the 'source' of the pressure can be considered to be the base of the foundation; the wider this is the greater will be the width

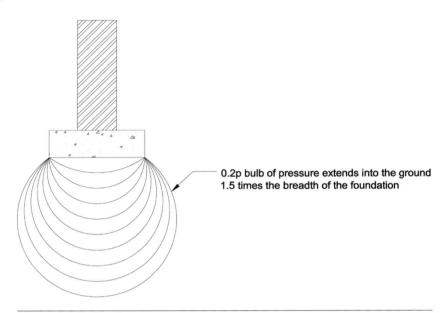

0.2p bulb of pressure extends into the ground 1.5 times the breadth of the foundation

Figure 5.1 Bulbs of pressure

and depth of the *bulbs of pressure* emanating from the source. It is generally regarded that pressures on the ground of less than one-fifth of those imparted at the surface will be negligible (i.e. if the surface pressure is 1p the least significant pressure will be 0.2p). This 0.2p pressure bulb will extend into the ground to a depth equivalent to one and a half times the width of the foundation (see Figure 5.1).

Over-stressing of the ground beneath is avoided by providing sufficient area of foundation. Consider footprints in the sand. A man weighing 90kgs with a heel size of 50mm x 50mm would impart a force of 0.35N/mm² on the sand. A woman weighing 60kgs but with a heel size of 5mm x 5mm would impart a force of 23.5N/mm² on the sand. It is clear from this that the woman's heel would make the deeper imprint into the sand, even though the load (i.e. the weight of the person) was less than that of the man. The important thing to consider here is not so much the load but the area that the load is being transferred to the ground over (i.e. the load per unit area).

Unequal settlement is avoided by ensuring that the centre of gravity of the applied load coincides with the centre of area of the foundation. If these two points do not coincide then the foundation will be unbalanced, causing it to tip and to over-stress the ground on one side of the foundation to a greater extent than on the other less loaded side.

Traditional strip foundations

These consist of a strip of concrete under a continuous wall that carries a uniformly distributed load (i.e. the load from the wall onto the foundations should be the same all the way along the length of the wall). The imposed load is therefore considered as a load per metre run of loadbearing wall.

The width of the strip of concrete in the foundation is related to the imposed load from the building, via the loadbearing wall, and the bearing capacity of the soil. The bearing capacity of the soil is considered to be the load the soil is capable of carrying without excessive settlement per unit area of foundation.

$$\text{Required width of foundation (metres)} = \frac{\text{Load per metre run of wall (kN/m)}}{\text{Bearing capacity per m}^2 \text{ area of soil (kN/m}^2)}$$

A factor of safety figure would normally be attached to the above calculation to take account of exceptional circumstances when the usual imposed loading from the building was increased. By taking this figure into account in the calculations the size of the foundation would always be large enough to transfer the exceptional load safely to the ground beneath. Exceptional loading may occur, for instance, where a party is being held in the house and the live loading from the people in the house increases substantially. A factor of safety figure of 2 effectively increases the total imposed load by a factor of 2 (i.e. doubles it) and thus doubles the width of the strip foundation for the same soil bearing capacity. However, for most domestic properties a factor of safety of 1.5 should be sufficient.

Self-assessment question 5.4

Calculate the width of a strip foundation for a house with a total imposed loading of 40kN/m run of loadbearing wall, constructed on soil having a safe bearing capacity of 120kN/m² and using a factor of safety of 1.5.

In cases of light loading on reasonably strong soils, a strip no wider than the width of the wall need be provided. In practice, however, some spread is usually provided to allow working room for the bricklayers building the walls (see Figure 5.2).

The Building Regulations Approved Document A, Section 2E3 (Table 10) stipulates minimum widths of strip foundations according to soil type and loads being carried, to ensure that the loads exerted on the soil are within its maximum bearing capacity. These minimum widths are often likely to be greater than those determined by the calculation method previously described. This is because the Building Regulations will consider the poorest loadbearing capacity for a particular soil in order to ensure that all foundations designed to bear onto that soil type would be adequate. Designers do not have to use the minimum width of foundation stipulated in the Building Regulations, but they must be able to provide calculations to show that the foundation that they have designed will safely carry the imposed load without failure.

However, where a mechanical excavator is being used to dig the trench for the foundation, the width of the trench and therefore the width of the foundation will often be dictated by the size of the bucket used on

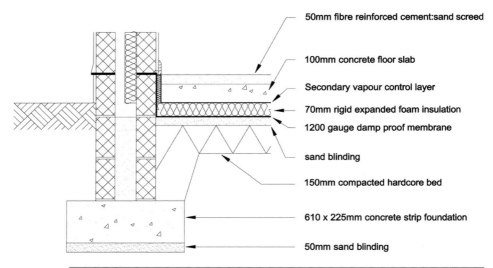

50mm fibre reinforced cement:sand screed

100mm concrete floor slab

Secondary vapour control layer

70mm rigid expanded foam insulation

1200 gauge damp proof membrane

sand blinding

150mm compacted hardcore bed

610 x 225mm concrete strip foundation

50mm sand blinding

Figure 5.2 Traditional strip foundations

the excavator. This is acceptable provided it does not produce a foundation width that is narrower than that determined by calculation or by the requirements suggested within the Building Regulations, where these are used to determine the width of the foundation.

Where the edges of the foundation project beyond the faces of the wall being supported, bending will occur in the concrete of the foundation as a result of the resistance produced by the soil to the imposed load that the foundation is transferring to the soil (see Figure 5.3).

If this bending is excessive, failure of the foundation due to shear stresses may result. To prevent this from occurring, the foundation should be constructed of an adequate thickness such that the line of shear failure (normally sloping down at an angle of 45⁰ from the base of the wall) does not affect the functioning of the foundation. Thus, in order to

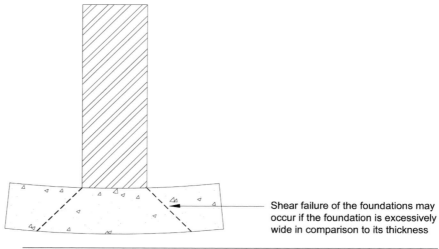

Shear failure of the foundations may occur if the foundation is excessively wide in comparison to its thickness

Figure 5.3 Lines of shear failure on wide and thin foundations

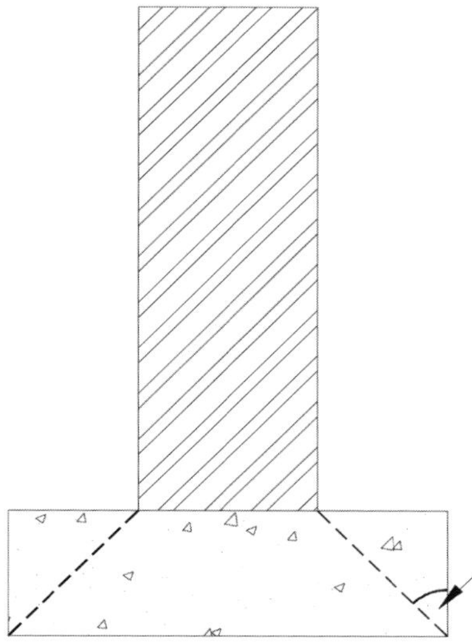

Angle should not be greater than 45° to overcome shear failure

Figure 5.4 Design of foundation to prevent shear foundations

keep the shear plane entirely within the width of the foundation the thickness of the foundation should be at least equal to the width of projection of the foundation from the base of the wall (see Figure 5.4).

Self-assessment question 5.5

What would be the minimum thickness of a strip foundation 600mm wide that is supporting a 300mm wall located centrally upon it?

The Building Regulations Approved Document A, Section 2E2 also stipulates that the minimum thickness of a strip foundation should be 150mm. This, of course, may be greater where the projection of the edge of the foundation from the base of the wall is greater than 150mm.

Wide strip foundations

Where strip foundations are too wide to economically increase their thickness in line with the '45⁰ rule', the lower part of the concrete foundation must be reinforced with high tensile steel reinforcement rods to combat the tensile stresses set up by the bending tendency induced by the soil's reaction to the imposed loading. The reinforcement rods need to be protected from corrosion by a suitable *cover* of concrete. These are known as wide strip foundations (see Figure 5.5).

For un-reinforced strip foundations a concrete mix of 50kgs of Portland cement to not more than 200kgs of fine aggregate and 400kgs of coarse aggregate is suitable. For reinforced foundations, however, a stronger mix should be provided. Concrete mix proportions will be considered in more detail in Chapter 6 (concrete). The foundation concrete should be well

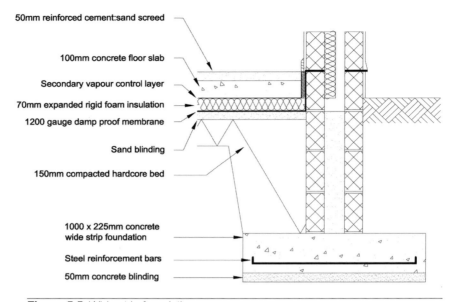

50mm reinforced cement:sand screed

100mm concrete floor slab

Secondary vapour control layer

70mm expanded rigid foam insulation

1200 gauge damp proof membrane

Sand blinding

150mm compacted hardcore bed

1000 x 225mm concrete
wide strip foundation

Steel reinforcement bars

50mm concrete blinding

Figure 5.5 Wide strip foundation

compacted in layers and a suitable 'blinding' layer of concrete 50mm in thickness should be laid on the base of the excavated trench, to even out any discrepancies in the level of the soil, before the foundations are laid.

Deep strip foundations

In firm clays the subsoil is capable of carrying substantial loads and may only require a foundation to be slightly wider than the wall it is supporting. The foundation still needs to be deep enough to overcome the problems of seasonal changes in the moisture content of the subsoil. A deep, narrow foundation, perhaps 350mm wide and 1.5m deep, can then be constructed (see Figure 5.6). These deep strip foundations are sometimes called trench fill foundations. Because the trench is so narrow it would be impossible to excavate the trench by hand or construct brickwork footings up to ground level within the trench. It is therefore easier to excavate the trench by machine and to fill the completed trench with concrete to within a short distance of the ground.

To enable this kind of foundation to be used, the soil must be firm enough to be self-supporting so that earthwork support systems can be dispensed with, and the amount of excavation must be extensive enough to make the use of a machine economic.

Self-assessment question 5.6

Apart from the firmness of the soil, why else could earthwork support systems be dispensed with in trenches for this type of foundation?

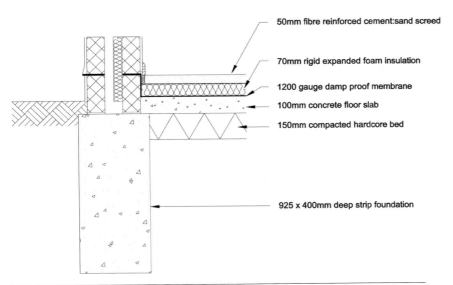

50mm fibre reinforced cement:sand screed

70mm rigid expanded foam insulation

1200 gauge damp proof membrane

100mm concrete floor slab

150mm compacted hardcore bed

925 x 400mm deep strip foundation

Figure 5.6 Deep strip foundation

Stepped foundations

It is important to ensure that the foundations to a building bear horizontally onto the ground. If they do not there is a possibility of them sliding when the ground beneath is lubricated by excess water. It is also difficult to prevent the applied load from over-stressing the 'leading edge' of the foundation where it is not on a level plane. On a sloping site, therefore, this will entail the foundations at the top of the slope being positioned deep in the ground in order to be level with those at the bottom of the slope, or the ground will need to be excavated to the same level over the entire area of the building. Both these methods would involve substantial extra excavation and would be expensive.

An alternative method is to step the foundations parallel to the slope of the ground. To prevent differential settlement in a stepped traditional strip foundation, the height of the steps should not exceed the thickness of the foundation (see Figure 5.7). At each step the higher foundation should overlap the lower foundation for a distance equal to the thickness of the foundation, or twice the height of the step, whichever is the greater, and should not be less that 300mm (Building Regulations Approved Document A, Section 2E2). For deep strip or trench fill foundations in these conditions the minimum overlap at the step should be twice the height of the step or 1m, whichever is the greater.

Self-assessment question 5.7

A traditional strip stepped foundation is being designed. The calculated size of the strip foundation is 550mm wide x 200mm thick. What would be the maximum height of each step and the minimum overlap of the higher foundation to the lower foundation at each step?

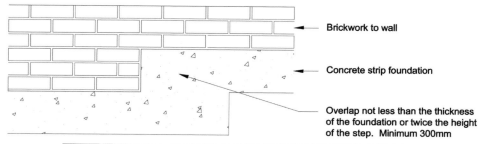

Brickwork to wall

Concrete strip foundation

Overlap not less than the thickness of the foundation or twice the height of the step. Minimum 300mm

Figure 5.7 Stepped foundations

Raft slab foundations

Where no firm soil exists at a reasonable depth below the surface then the width of the strip foundations becomes so large that they combine. The foundation thus becomes a large reinforced concrete slab covering the whole area of the building and constructed just below the surface of the ground. The foundation works on the principle that by widening the area that the load is being imposed onto the ground, the load per unit area will be reduced to a level that can safely be supported by the ground beneath (refer to the example of the footprints in the sand considered earlier).

The slab produced may be used as a ground floor slab and is lightly reinforced with fabric mesh reinforcement to counteract stresses formed within the foundation by the transfer of loads to the subsoil and the soils' reaction to those loads. The slab is thickened beneath areas where loads are imposed (i.e. loadbearing walls) and may resemble a traditional strip or deep strip foundation at these points (see Figures 5.8 and 5.9). The reinforcement in these areas needs to be in the lower part of the slab, where the tensile stresses are greatest. However, between these points of loading, the stress distribution within the slab is reversed and the tensile stresses are greatest near to the top of the slab, necessitating the location of the main reinforcement at this point.

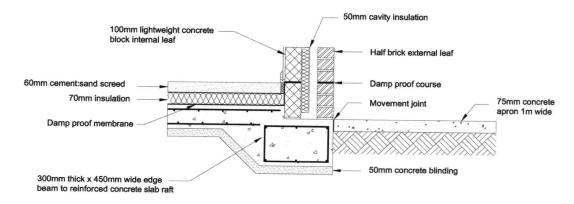

100mm lightweight concrete block internal leaf

50mm cavity insulation

Half brick external leaf

60mm cement:sand screed

70mm insulation

Damp proof course

Movement joint

75mm concrete apron 1m wide

Damp proof membrane

300mm thick x 450mm wide edge beam to reinforced concrete slab raft

50mm concrete blinding

Figure 5.8 Raft slab foundation with shallow edge beam

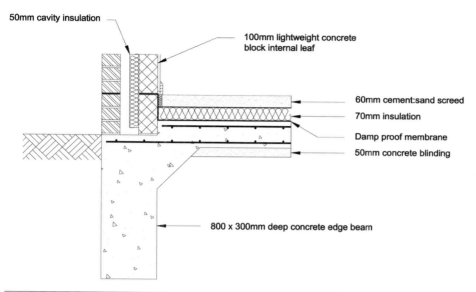

50mm cavity insulation

100mm lightweight concrete block internal leaf

60mm cement:sand screed

70mm insulation

Damp proof membrane

50mm concrete blinding

800 x 300mm deep concrete edge beam

Figure 5.9 Raft slab foundation with deep edge beam

At the edge of the building the raft slab should be extended to protect the soil beneath the external walls from possible frost action. This 'apron' may be just 300mm wide in sandy soils, but as wide as 1.5m in clay soils (see Figure 5.8).

Self-assessment question 5.8

Why is the soil at the edge of the building beneath a raft slab foundation vulnerable to frost action?

Deep foundations

These are used where firm soils do not exist at a level close to the surface of the ground but do exist at a much lower depth. Thus deep foundations are used to transfer the loads of the building through the poor loadbearing strata down to the higher loadbearing strata beneath. Obviously the deeper the foundation has to be taken in order to reach this higher loadbearing stratum the higher will be the cost of installing the foundation. For domestic construction, therefore, the maximum depth of a deep foundation will be approximately 4m. A short bored pile foundation is suitable.

This consists of a column of concrete, poured within a cylindrical shaft bored by an auger and extending to the higher loadbearing capacity subsoil. The auger may take the form of a large helically-shaped drill, mounted on the back of a lorry or tractor. Alternatively it may be a cylinder with a cutting edge that can be dropped from a small tripod stand. The weight of the cylinder allows a quantity of soil to be cut

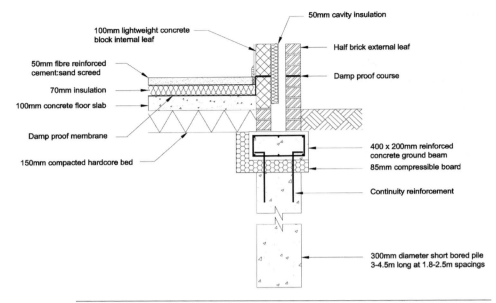

100mm lightweight concrete block internal leaf

50mm cavity insulation

Half brick external leaf

50mm fibre reinforced cement:sand screed

Damp proof course

70mm insulation

100mm concrete floor slab

Damp proof membrane

150mm compacted hardcore bed

400 x 200mm reinforced concrete ground beam

85mm compressible board

Continuity reinforcement

300mm diameter short bored pile 3–4.5m long at 1.8–2.5m spacings

Figure 5.10 Short-bored pile foundation

forming the shaft. This cut soil can be removed when the cylindrical cutter is raised to the surface. This technique is useful where space is particularly limited for forming the pile shaft.

A reinforced concrete beam supports the loadbearing wall over the piles and links all the piles together. This ground beam is reinforced in its lower part and, in order to maintain continuity in the transfer of loading from the ground beam into the piles, the reinforcement is also embedded in the top section of each pile. The piles are normally 250–350mm in diameter, between 2.0–4.0m in length and spaced at 1.8–2.5m centres or more frequently at points of high load (such as under fireplace flues and at reveals of windows and doors) and at corners of buildings (see Figure 5.10).

Little excavation is performed with this type of foundation, and in ground where a high water table exists and trench excavation is difficult, this type of foundation may be considered ideal.

Self-assessment question 5.9

If the higher loadbearing stratum is deeper than 4m, what would be the most appropriate choice of foundation?

An alternative system is also available that combines short bored or driven pile foundations with a reinforced concrete slab, stiffened at its edges by a reinforced concrete beam.

The effect of trees on foundations

The roots of plants take water from the soil during periods of low rainfall. This may occur up to a depth of 5m beneath large trees and cause soil shrinkage of up to 100mm measured at the surface of the ground.

A building constructed on shallow foundations close to existing trees may be liable to seasonal movement or even long-term settlement dependent on root growth. Even where construction occurs on sites where there are no existing trees, any new trees that are planted may also affect the foundations of the houses located nearby.

As a rough guide, buildings constructed on shallow foundations should not be located closer to trees than the mature height of that tree. New trees should also not be planted adjacent to buildings inside this distance (Building Research Establishment Digests 240 & 241).

The problem is more pronounced in shrinkable clay soils, which are often found in the south-east corner of England. Where shrinkable clay soils exist and there is the possibility that tree roots could cause excessive shrinkage of the soil around the foundations during periods of dry weather, a short bored pile foundation may be advisable. In shrinkable clay soils where the Plasticity Index (see the previous chapter on soil testing) is greater than or equal to 10 per cent, strip foundations need to be taken to a depth where the anticipated ground movements will not impair the stability of the building.

The Building Regulations Approved Document A Section 2E4 recommends that the depth to the underside of the foundations on clay soils should not be less than 750mm, although this depth will commonly need to be increased in order to transfer the loading from the building onto satisfactory ground. In areas where shrinkable clay subsoils are not a significant problem, but the proximity of nearby trees is a concern, the foundations may be of the deep strip variety. Alternatively a trench of width similar to that used for traditional strip foundations but filled with concrete and called a trench fill foundation may be utilised. The depth of these foundations will be dependent on factors such as the type of soil and the likely water demand of the nearby trees. Figures for these depths of foundation are available, published by the National House Building Council. Generally though, if a foundation depth greater than 2m is required, it is likely that a short bored pile foundation will be chosen rather than a deep strip or trench fill foundation.

When trees are felled the clay soil should be given sufficient time to regain the water lost by the action of the roots, so that subsequent swelling will not lift the foundations of the new building. This swelling can be in the order of 6mm per year over a period of ten years. To prevent this swelling action from damaging the foundations, the ground beams of short bored pile foundations and the sides of deep strip or trench fill foundations should be surrounded with a compressible material to absorb the stresses created by the swelling of the soil. Traditionally this compressible material has been ash or similar, however a modern alternative is clayboard.

Visit the companion website to test your understanding of Chapter 5 with a multiple choice questionnaire, and to see the following features illustrated with full-colour photos:

- **5.1** Cavity wall footings with concrete infill
- **5.2** Concrete blocks laid in footings
- **5.3** Concrete foundation
- **5.4** Concrete trench blocks for footings
- **5.5** Engineering bricks on top of concrete brick footings
- **5.6** Stepped foundation
- **5.7** Trench blocks being laid in a foundation trench
- **5.8** Trench fill foundation
- **5.9** Trench foundation for an internal wall

6

Concrete

Introduction

Concrete is a mixture of cement, fine aggregates such as sand, coarse aggregates such as gravel, crushed rock or lightweight granular materials, and water, combined together in the correct proportions to give a strong, dense, homogeneous material which can be easily moulded to desired shapes and is ideally suited for the manufacture of structural components.

The cement provides the setting and hardening element to the concrete. The water performs two main functions. Firstly it allows the cement to *hydrate*, which is the chemical reaction between cement and water allowing the cement to set and harden. Secondly, water is also included in a concrete mix to provide *workability* to the mix. This enables the constituent materials to be mixed together easily and to be placed in position and compacted easily. The coarse aggregates provide bulk to the concrete mix and also contribute towards the final strength of the concrete. The fine aggregates fill in the spaces between the coarse aggregate particles, making the concrete denser and less porous. They also combine with the cement and water to produce a *grout* that can provide a smooth surface finish to the concrete when it has set.

In this chapter, consideration will be given to the functional requirements of concrete, the types of cement, the types of aggregates, both coarse and fine, dense and lightweight, the design of concrete mixes, the achievement of workability, the water to cement ratio, the aggregate to cement ratio and the grading of aggregates.

Functional requirements

Concrete has to achieve four main functional requirements:

- Strength – If the concrete is to be used as a structural material it must be strong enough to resist the stresses imposed upon it by the load. The strength of the concrete is normally influenced by the proportioning of the materials and the type of coarse aggregate used. Concrete has high compressive strength but is relatively weak in tension. Therefore where concrete structural components need to be strong in both compression and tension, steel reinforcement bars or fabric mesh is added to the concrete member in positions where tensile forces are likely to be concentrated when the member is loaded. Steel has good tensile strength and the combination of concrete and steel in a *reinforced concrete* member will enable that member to support heavy structural loads.

- Durability – Where concrete is exposed to the external climate it will need to be weather resistant and frost resistant. In these situations the porosity of the concrete will need to be minimised. This can be adjusted by the proportioning of materials within the mix, the grading and type of aggregates in the mix, the amount of water used in the mix and the degree of compaction of the concrete when it is placed. Where concrete is used in structures that are buried in the ground it is necessary to consider whether materials present in the ground, such as sulphate salts, may have a detrimental effect on the concrete. Where high amounts of sulphate salts exist it is necessary to consider the selection of a cement with sulphate resisting properties for the concrete. Where steel reinforcement is used in reinforced concrete members it will need to be protected from corrosion by providing a *cover* of concrete. The amount of cover required will depend on the location of the concrete member.

- Fire resistance – Concrete has good fire resisting properties. Where steel reinforcement is used in reinforced concrete members then it will need to be protected by a cover of concrete. The amount of cover required will be dependent on the amount of fire resistance required. The fire resistance of concrete can be enhanced by the use of aggregates such as blastfurnace slag, crushed bricks or lightweight aggregates.

- Thermal insulation – The thermal insulation capabilities of concrete can be improved by the use of lightweight aggregates instead of dense aggregates. These aggregates contain more air, which is a good thermal insulator. Other ways in which more air can be introduced to the concrete is by the use of *air entrainment*. This can be achieved by adding a material to the concrete mix that will react with the water in the mix and produce a number of air bubbles that will be 'locked in' to the concrete when it has set. Increasing the thermal insulation capabilities of a concrete mix is also likely to increase the porosity of the concrete. This may not be desirable in some situations.

The achievement of these properties depends upon the choice and proportioning of the constituent materials in the mix, as well as good workmanship in mixing. Not all of the above properties may be required by one particular concrete.

Sustainability – cement has a high amount of embodied energy in its manufacture although the amount of CO_2 produced in the manufacturing process is being reduced through the burning of alternative fuels instead of fossil fuels and the use in some cements of proportions of fly ash or granulated blastfurnace slag, both waste products of other industries. However the energy used in its transportation is relatively low as little cement is imported. Aggregates make up a substantial proportion of a concrete mix and are therefore important when considering the sustainability of the final material. Dense aggregates, such as gravels, incur embodied energy only through their extraction. However the extraction process can destroy natural habitats and deface the landscape. Aggregates obtained from crushed rock also incur embodied energy in the extraction process and further energy in the crushing process. Both these materials have relatively low transportation energy requirements. There are a wide variety of materials used for lightweight aggregates. Many of these materials derive embodied energy through the manufacturing process. Some are natural products but have high transportation energy costs, as they are imported. However, like cement, some lightweight aggregates are manufactured from fly ash or granulated blastfurnace slag and therefore have a high recycled content, reducing the amount of these materials being sent to landfill.

Cements

These provide the setting and hardening element to the concrete. Cement provides strength and is a hydraulic material; that is, it requires the presence of water, not air, in order to hydrate.

There are two parts to the reaction between cement and water. An initial chemical set is followed by a long hydration period, in which the cement gains its final strength. Heat is produced during the setting process and this may need to be controlled if cracking is to be prevented.

Cements are manufactured in three main types:

- Portland cements
- supersulphated cements
- high alumina cements.

Portland cements

These are manufactured by heating of a slurry of clay containing silica, alumina and iron oxide, with limestone in a rotating furnace. This process changes the limestone into lime (called calcification). The resulting clinker is then ground with a small proportion of gypsum to aid setting. Most Portland cements are manufactured to the requirements of *BS EN 197-1:2000 Cement composition, specifications and conformity criteria for common cements*.

There are four main chemical compounds in Portland cements, which are shown in Table 6.1.

Portland cement is the most commonly used of the three main types and it is available in a variety of sub types for different end uses:

- *Ordinary Portland Cement (OPC)* – this is the most common sub type. Its setting time and strength development is good enough for most uses.
- *Rapid Hardening Portland Cement (RHPC)* – this develops strength more rapidly than OPC and therefore earlier striking of *formwork* is

Table 6.1 Constituent materials of Portland cements

Chemical name	Empirical formula	Shorthand formula
Tricalcium Silicate	Ca_3SiO_5	C_3S
Dicalcium Silicate	Ca_2SiO_4	C_2S
Tricalcium Aluminate	$Ca_3Al_2O_6$	C_3A
Calcium Aluminoferrite	$2Ca_2AlFeO_5$	C_4AF

possible. The cement is similar in composition to OPC but it is more finely ground. This does not affect the setting time, however, so concretes made with this cement have similar periods in which they remain workable as do concretes made with OPC.

- *Ultra Rapid Hardening Portland Cement* – this is very finely ground cement (even more finely ground than RHPC) with a high proportion of gypsum. The initial development of strength is great.

Self-assessment question 6.1

Why should finer grinding of the cement improve the initial development of strength in a concrete?

- *Sulphate Resisting Portland Cement (SRPC)* – this cement has a modified chemical composition whereby the Tricalcium Aluminate (C_3A) content is limited. It is the hydration product of this compound in most cements that is susceptible to attack by sulphate salts, which may be present in high concentrations in some soils. The sulphate salts in the soil react with the hydrated Tricalcium Aluminate present in the hardened cement within the concrete, causing the growth of a crystalline mineral that causes the cement within the concrete to expand. This expansion can ultimately disrupt the concrete, causing it to crumble. It is manufactured to the requirements of *BS 4027:1996 Specification for sulphate resisting Portland cement*.
- *Portland Blastfurnace Cement* – this is made by grinding a mixture of granulated blastfurnace slag with OPC clinker. It has lower strength development than OPC but better sulphate resistance (though not as good as SRPC) and is often specified when large masses are to be concreted. Its sustainability credentials are also better than OPC. It is manufactured to the requirements of *BS EN 197-4:2004 Cement composition, specifications and conformity criteria for low early strength blastfurnace cements*.

Self-assessment question 6.2

Why should the strength development of concrete using Portland Blastfurnace Cement be slower than that for concrete using OPC?

- *Low Heat Portland Cement* – this cement contains less Tricalcium Silicate (C_3S) and more Dicalcium Silicate (C_2S) than OPC. This slows down the rate of hydration and, consequently, the rate of strength development. Again it is often specified where large masses are to be concreted. It is manufactured to the requirements of *BS 1370:1979 Specification for low heat Portland cement*. A Low Heat Portland Blastfurnace Cement is also available which combines the features of Portland Blastfurnace Cement and Low Heat Portland Cement. It is manufactured to the requirements of *BS 146:2002 Specification for blastfurnace cements with strength properties outside the scope of BS EN 197-1*.

Self-assessment question 6.3

Why should cements with a slower rate of early strength development be more suitable for concrete used in mass structures than cements with a high rate of early strength development?

- *White Portland Cement* – this is made by carefully selecting raw materials such as china clay and high grade chalk, which are free from colour forming impurities such as iron oxide. The heat used in the manufacturing process is also closely controlled. The resulting cement is primarily used where white concrete is required to provide a visual effect.
- *Coloured Portland Cements* – these are based on White or Ordinary Portland Cement with 5–10 per cent coloured pigments. The pigment dilutes the active cement in a specific volume, thus a greater cement content is required in a mix to achieve the required final strength.
- *Water repellent cement* – special additives are added to OPC to make this cement less permeable to water and this property is encompassed in the resulting concrete. However good mix control is important for this to be successfully achieved.

Self-assessment question 6.4

Why should good mix control be important when using water repellent cement in a waterproof concrete mix?

- *Hydrophobic Cement* – this cement has been developed to prevent partial hydration of cement during storage in humid conditions (so called 'air setting'). Additives form a water repellent film around each particle, which is rubbed off in the mixing process.

Supersulphated cement

This cement is made by grinding together granulated blastfurnace slag with calcium sulphate and Portland cement clinker approximately in the

Table 6.2 Minimum compressive strength in N/mm² for Portland cements

Type of Portland cement	3 days	7 days	28 days
Ordinary	13	-	29
Rapid hardening	18	-	33
Sulphate resisting	8	14	-
Low heat	5	-	19
Blastfurnace	8	14	22
Low heat blastfurnace	3	7	14

proportions of 17:2:1 as specified in *BS EN 15743:2010 Supersulphated cement. Composition, specifications and conformity criteria*. It has a very low C_3A content and thus has resistance to sulphate attack. It also has high resistance to acid and alkali attack. As would be expected its development of early strength is slow and thus the rate of heat evolution is also slow. This can have a significant effect in cold weather and in these conditions low temperature steam curing may be advantageous.

High alumina cement (calcium aluminate cement)

This cement is manufactured by fusing together limestone and bauxite in reverberatory furnaces. The 'pigs' thus produced are broken and ground to produce the resultant cement as specified in *BS EN 14647:2005 Calcium aluminate cement. Composition, specifications and conformity criteria*. It develops an extremely high early strength, which allows for early loading, but the rate of heat evolution is also high and this may need to be checked by cooling the concrete during curing. Concrete incorporating high alumina cement has high sulphate resistance, high resistance to acids (but not alkalis) and good abrasion and impact resistance but is approximately three times more expensive than OPC.

Following the reporting, in the early 1970s, of structural failures amongst prestressed concrete members which used high alumina cement in their manufacture, it was discovered that when concrete using high alumina cement reaches temperatures of around 30°C in moist conditions, the hydrates which are present in the cement transform from a hexagonal structure to a cubic structure. As this change to the crystalline structure within the cement occurs, water is liberated. This water reacts with carbon dioxide present in the atmosphere and frees alkalis present in the cement to form alkali carbonates. These then attack and decompose the calcium aluminate hydrates that are the main bonding agents in the concrete. The reaction can continue until the mechanical properties of the concrete have been virtually destroyed.

This problem can be overcome by proper mix design and curing of the concrete, but the results of incorrect mix design and curing can be so catastrophic that high alumina cement is no longer specified for structural concrete.

Supersulphated and high alumina cements are only specified when their special characteristics are particularly desirable in the finished concrete. They must not be mixed together or mixed with any other cement.

Self-assessment question 6.5

Suggest a suitable cement that could be used in concrete for a mass foundation in soil that has a SO_3 content of 0.9% (i.e. 90g of Sulphur Trioxide per litre of ground water) and a pH value of 4.5.

Aggregates

Although it is quite possible to produce a concrete that is simply a mixture of cement and water, the resulting material would be extremely expensive and suffer from quite high amounts of drying shrinkage, coupled with poor durability. To reduce these adverse features, organic matter in the form of aggregates is added to the concrete mixture.

These aggregates make up a substantial proportion of a concrete mix and should not react adversely with the cement, nor should they become unstable when subjected to changes in their moisture content. The shape and surface texture of the aggregate particles and their grading are important factors that will influence the workability and final strength of the concrete.

Self-assessment question 6.6

What is meant by the term 'grading' when applied to aggregates and why should it be important?

Aggregates fall into two main categories related to their particle size; fine aggregates made up of material that will pass a BS 5mm sieve (i.e. the particles are smaller than 5mm in diameter) and coarse aggregates, made up of larger particles. The coarse aggregates generally contribute to the strength of the finished concrete, whilst the fine aggregates help to create a smooth surface finish.

Aggregates may also be classified by their density. Dense aggregates are those that have a density greater than 960kg/m^3 for coarse aggregates and 1200kg/m^3 for fine aggregates. Aggregates with densities below these values are classified as lightweight aggregates.

Dense aggregates

These are usually obtained from natural sources such as washed gravels and crushed rocks, graded according to limits specified by *BS EN*

12620:2002 Aggregates for concrete and Amendment A1:2008. Washed sands for fine aggregates are graded in zones according to particle size. Zone 1 is the coarsest and zone 4 is the finest. Natural sands tend to fall into zones 3 and 4 whilst sands derived from crushed rock tend to fall into zones 1 and 2.

Air cooled blastfurnace slag coarse aggregate is a by-product from the manufacture of pig iron and is covered by *BS EN 12620*. It gives similar strength characteristics to the concrete as other dense aggregates but also provides good fire resistance. Crushed brick aggregate also has good fire resistance.

Self-assessment question 6.7

Why should blastfurnace slag and crushed brick aggregates have better fire resistance than natural gravel aggregates?

Lightweight aggregates

There are several types of lightweight aggregate in common use, each having its own characteristics of particle shape and surface texture. They all tend to be cellular in structure and it is the presence of this air that reduces their density.

Lightweight aggregate concretes give better fire resistance and thermal insulation values than most concretes manufactured from natural dense aggregates. They do, however, have a high drying shrinkage and are therefore more susceptible to shrinkage cracking than their denser counterparts.

Self-assessment question 6.8

Why should lightweight aggregate concretes have a higher drying shrinkage than dense aggregate concretes?

The angularity and rough surface texture of the lightweight aggregate particles also leads to poorer workability of the concrete mix.

The higher porosity of lightweight aggregate concretes tends to retard their natural weather resistance and durability.

Lightweight aggregate concretes generally have lower crushing strengths than their denser aggregate counterparts.

The types of lightweight aggregate in common use are:

- *Furnace clinker* – these are well-burnt residues from furnace fuels that have been fused or sintered into lumps. *BS EN 13055-1:2002 Lightweight aggregate. Lightweight aggregates for concrete, mortar and grout* limits the combustible content of clinker in this material for different types of concrete, along with its sulphate content.
- *Foamed blastfurnace slag* – this is molten slag that has been treated with water to produce a cellular dry product, which is then crushed

and graded to requirements. *BS EN 13055-1* specifies maximum densities of 800 kg/m³ for coarse aggregate and 1120kg/m³ for fine aggregate.

- *Exfoliated vermiculite* – vermiculite resembles Mica and consists of many layers of flakes. These flakes open out, or exfoliate, when the aggregate particles are heated to a high temperature. *BS EN 13055-1* specifies a maximum density of 130kg/m³.
- *Expanded perlite* – this is a volcanic rock that expands to form a cellular material when heated. *BS EN 13055-1* specifies a maximum density of 240kg/m³.
- *Pumice* – this is cooled volcanic lava that is crushed, washed and graded. *BS EN 13055-1* specifies a maximum density of 960kg/m³ for coarse aggregate and 1200kg/m³ for fine aggregate.
- *Expanded clay and shale* – certain clays and shales expand, on heating, to form a cellular material. The process is carried out in rotary kilns so the aggregates that are formed resemble rounded pellets. *BS EN 13055-1* specifies a maximum density of 960kg/m³ for coarse aggregate and 1200kg/m³ for fine aggregate.
- *Sintered pulverised fuel ash* – pulverised fuel ash (PFA) is a residue from the burning of powdered coal in coal fired power stations. The ash is moistened, formed into pellets and then fired at high temperatures to form hard cellular pellets, which are then crushed and graded. *BS EN 13055-1* specifies a maximum density of 960kg/m³ for coarse aggregate and 1200kg/m³ for fine aggregate.

The design of concrete mixes

General principles

Concrete must be strong enough when hardened to resist the stresses to which it is to be subjected. It must also be able to withstand the action of weather if it is to be used externally. In addition, when freshly mixed, it must be easily handled without *segregation* of the aggregates and be easily compacted in position. In general, the more it is compacted the greater will be its resultant strength and durability. These properties are also affected by the water:cement ratio of the concrete.

To obtain a dense concrete, the air voids in the finished product should be a minimum. This is achieved by proportioning the aggregates, cement and water, such that the mix is sufficiently workable to be properly compacted. Excess water in the mix will lead to a weaker and less durable concrete, since the excess water will evaporate during the drying process, leaving air voids in the concrete.

Self-assessment question 6.9

What is meant by the term 'workability' in concrete mix design?

Normally there is approximately twice the quantity of coarse aggregates to fine aggregates in a concrete mix. The overall grading of the aggregate has an effect on the amount of water added since 'fine' gradings (i.e. a greater proportion of fine aggregates to coarse aggregates than normal) require more water than 'coarse' gradings (i.e. a greater proportion of coarse aggregates to fine aggregates than normal) to achieve the same degree of workability.

Self-assessment question 6.10

Why should fine gradings require more water than coarse gradings to achieve the same degree of workability?

Concrete mixes are usually specified by the proportions of materials they contain (i.e. 1 part cement to 2 parts fine aggregate to 4 parts coarse aggregate) or in terms of their compressive strength required at a particular age (i.e. $30N/mm^2$ at 28 days). Proportions are more correctly specified by weight than by volume.

Self-assessment question 6.11

Why are proportions of materials in a concrete mix more correctly specified by weight than by volume?

The water:cement ratio

This is the volume of water in a mix, expressed as a ratio to the weight of cement. A water:cement ratio of 0.45 would indicate that 0.45 litres of water were added for every 1kg of cement in the mix (this is based on the premise that 1 litre of water weighs 1kg).

The total weight of water includes the moisture contained in the aggregates as well as the 'free' water added to the mix.

Self-assessment question 6.12

How can the amount of moisture contained in the aggregates be determined?

A certain proportion of the water combines with the cement and sets up the chemical reaction called hydration, which sets and hardens the concrete. The remainder of the water added to the mix is required to make the mix workable.

The amount of water used should be the minimum necessary to give sufficient workability for efficient compaction and to reduce the porosity of the hardened concrete. Where the durability of the concrete is particularly important, the water:cement ratio may need to be limited.

Workability

The greatest degree of compaction can be achieved when the workability of the mix is such that it suits both the type of concrete member being formed and the method used to compact the concrete. A less workable (stiffer) mix can be used when compacting by vibrator than that used when compacting by hand.

Workability depends upon:

- the water:cement ratio
- the proportion of cement to aggregate. For the same water:cement ratio a rich mix (i.e. a mix with a greater proportion of cement to aggregates than normal) is more workable than a lean mix (i.e. a mix with a lesser proportion of cement to aggregates than normal)
- the type, maximum size and grading of the aggregate.

A good mix therefore has a water:cement ratio giving the strength and durability required and is sufficiently workable to be properly compacted with the means available. In designing a mix the water:cement ratio is chosen to give the required strength and durability, whilst the ratio of cement to aggregate and the grading of the aggregate determine the required workability.

The aggregate:cement ratio

When the degree of workability has been fixed, the proportion of cement to the total aggregates necessary to obtain such workability must be determined. Proportions of fine and coarse aggregate will vary according to the type and grading of each, and according to the workability and cement content of the concrete.

An excessive proportion of fine aggregate requires a high water:cement ratio (see self-assessment question 6.10), which increases the shrinkage and reduces the strength, in order to give good workability.

The grading of aggregates

The proportion of fine to coarse aggregates should be kept as low as possible, but should not be so low as to cause segregation or to make it difficult to obtain a good surface finish. The same workability and strength can be provided by using a smaller proportion of fine sand or a larger proportion of coarse sand.

Visit the companion website to test your understanding of Chapter 6 with a multiple choice questionnaire.

CHAPTER 7

Bricks and blocks

Introduction

The basic components of construction for walls are bricks, blocks and stones. Because blocks are more porous than bricks and therefore not as weather resistant, and because stones tend to be more expensive than bricks for house wall construction in most parts of the United Kingdom, it is usual to select bricks for external wall construction and blocks for internal wall construction. Cavity walls generally have a brick external leaf and a block internal leaf. This chapter will consider the main materials used for brick manufacture and compare their physical characteristics. Similarly the main materials and types of block will also be considered along with their physical characteristics. Building stones will not be considered.

Bricks

Bricks are manufactured from three basic raw materials:

- clay
- calcium silicate
- concrete.

There are a large number of brick manufacturers making a large number of different types of brick. These may be classified in a number of ways.

- Classification by use – common bricks have acceptable strength, water absorption, durability, thermal and moisture movement, thermal conductivity, and sound insulation performance characteristics, but they are cheaply produced and their appearance is poor. They can therefore be specified in locations where they will not be seen in the finished wall. Sometimes this may mean that the wall is constructed from common bricks and is then covered with an external and internal facing material such as render and plaster. Facing bricks, on the other hand, are more expensive to produce and their appearance is much better and they can therefore be used in positions where they will be seen. Their performance characteristics in all other respects are similar to those for common bricks. Engineering bricks, as the name suggests, are stronger than common or facing bricks and have lower moisture absorption characteristics. They too are more expensive than common bricks and are specified for use where their enhanced strength and low porosity are important, for example below ground in footings. Their appearance is not particularly appealing and they would therefore tend to be used in areas where they cannot be seen.

- Classification by method of manufacture – clay, calcium silicate and concrete bricks are made in entirely different ways from each other. The shape of a clay brick may be formed in a wooden mould, extruded through a metal die or pressed between steel plates. Each method of forming the bricks will have an influence on its cost, appearance and, to a certain extent, its physical characteristics.
- Classification by structure – during the formation of the bricks, material is often removed from the internal part of clay bricks to reduce their weight and to improve firing in the kiln. Removal of material from the structure of the brick can be achieved by forming perforations through the centre of the brick, forming a cellular structure in the centre of the brick or pressing a *frog* in the shape of a prism in the top surface of the brick. Calcium silicate bricks tend to be solid or have a slight frog depression.
- Classification by finish – facing bricks can have their appearance enhanced by finishing their faces with sand or sawdust. They can also have their faces textured by combing or hammering the surface before firing. Facing bricks may also have different colours either through natural pigments present in the clay or pigments added during the manufacturing process. Common bricks, engineering bricks, calcium silicate bricks and concrete bricks tend to have a smooth finish.

Brick terminology

BS EN 772-16:2011 Methods of test for masonry units. Determination of dimensions specifies a brick as being a masonry unit having work sizes not exceeding 337.5mm in length or 112.5mm in height. A standard format size of a brick is normally 215mm long x 102.5mm wide x 65mm high. These dimensions relate closely to the old imperial measurements for a brick. There have been attempts to introduce bricks having metric dimensions, but these have not been popular, as the appearance of walls constructed from them differs from that of walls built from the traditionally sized bricks. Metric modular bricks are available in sizes 190mm long x 90mm wide and either 65 or 90mm high or 290mm long x 90mm wide and 65 or 90mm high, but these account for only a very small proportion of the market.

The long face of the brick is termed the *header face*; the end face of the brick is termed the *stretcher face*. If a *frog* is used it is usually formed in the upper surface of the brick. The lower surface is termed the *bed face* (see Figure 7.1).

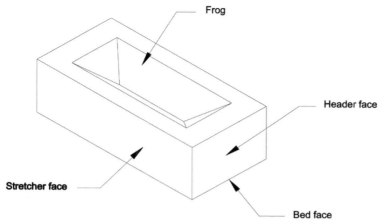

Figure 7.1 Brick terminology

The manufacture of clay bricks

The manufacture of clay bricks is covered by *BS EN 771-1:2003 Specification for masonry units. Clay masonry units*.

There are three basic stages in the manufacture of clay bricks:

- excavation and preparation of the raw material
- forming the shape required
- drying and firing the brick.

Excavation and preparation of the material

Clays used in brick manufacture are composed mainly of silica and alumina, together with impurities that might include natural pigments or even some natural fuels that are consumed during the firing process. The degree of impurities in the clay used for brick manufacture, particularly the amount of soluble salts, must be within acceptable limits set by *BS EN 772-5:2001: Methods of test for masonry units. Determination of the active soluble salts content of masonry units*.

The clay is quarried from the ground using a *dragline*. This machine is based on a crawler mounted mobile crane chassis with a large cable operated drag bucket attached to the jib. The bucket is thrown out to its furthest extremity by simultaneously slewing the jib and releasing the support cable. Being dragged back towards the machine by its drag cable then fills it.

After quarrying, the clay is crushed, screened to remove stones and other debris, mixed with water and then passed through a *pug mill* to produce uniform clay with high plasticity. The high plasticity created by the addition of the water helps in the moulding process.

Self-assessment question 7.1

Why is water added to the clay?

The air that has been introduced into the clay during the mixing process must then be removed by means of a vacuum pump and the clay may then be left to age to ensure it achieves a uniform distribution of water.

Self-assessment question 7.2

Why does the air need to be removed?

Forming the shape required

The bricks may be moulded by hand or by machine. Handmade bricks are formed by throwing a lump of clay into timber moulds that have previously been lined with sand or sawdust. This process produces relatively irregular shapes and dimensions that are sought by specifiers who do not want to use bricks having regular shapes and sizes for their buildings. The clay must be soft and have high plasticity for this process. The process is relatively slow and the bricks are therefore more expensive than those produced by machine.

Alternatively the clay may be forced through a die of the shape and dimensions of the base of the brick. The extruded clay is then cut into units corresponding to the brick height and length by stretched wires, in a similar manner to that used for cutting cheese. The clay needs to be of moderately stiff consistency. During this process perforations can be formed in the centre of the bricks to save clay and reduce weight. These perforations also improve the drying and firing of the bricks.

Stiff clays can be formed into bricks by mechanical pressing. This process allows the formation of a depression or frog in the upper face of the brick.

Drying and firing the brick

The 'green' bricks are fired in furnaces in which either the fire rotates (the Hoffman kiln) or the bricks are loaded onto trolleys and move through a stationary fire (the tunnel kiln). The temperature of firing is usually around $900^{\circ}C$, although engineering bricks are generally fired at higher temperatures.

Before being fired the bricks must be dried, otherwise they may crack when they are subjected to the high temperatures within the kiln. Leaving the bricks for a day or two at the entrance to the kiln before they are loaded into the kiln enables the outgoing hot gases from the kiln to dry the bricks.

After firing, the bricks are cooled at the exit to the kiln by the incoming air required by the kiln for the combustion process. This again helps to prevent cracking to the bricks caused by thermal shock through too rapid cooling in the open air.

It is important to ensure that the bricks are thoroughly fired in the kiln. Under-fired bricks are less durable than fully fired bricks. The amount of drying and firing that a brick requires is dependent to a large extent on its method of manufacture. Pressed bricks require the least amount of firing whilst handmade bricks require the most.

Self-assessment question 7.3

Why is the amount of drying and firing that a brick requires dependent on its method of manufacture?

The manufacture of calcium silicate bricks

These bricks, often known as *sandlimes* and *flintlimes*, are manufactured from a combination of sand or flint mixed with quicklime or hydrated lime and water in the proportion of 10 parts of sand or flint to 1 part of quicklime or hydrated lime to *BS EN 771-2:2011 Specification for masonry units. Calcium silicate masonry units*.

After mixing the mixture is pressed in steel moulds and then *autoclaved* for up to 12 hours in steam ovens at a temperature of 170°C and a pressure of approximately 10 atmospheres. This process enables the lime to react with the particles of silica, leaving no free lime.

The calcium silicate thus formed reacts with carbon dioxide when it is exposed to the air, forming calcium carbonate. This allows the brick to gain in strength and hardness.

The manufacture of concrete bricks

These are manufactured from concrete having a dense aggregate to *BS EN 771-3:2011 Specification for masonry units. Aggregate concrete masonry units (dense and lightweight aggregates)*. The mixture is placed into steel moulds and autoclaved with high pressure steam in a similar manner to that of calcium silicate bricks.

The bricks may be solid, perforated, hollow or cellular in structure.

Properties of bricks

The following physical characteristics will be considered for each of the three types of bricks:

- compressive strength
- water absorption
- durability
- thermal and moisture movement
- thermal conductivity
- sound insulation
- fire resistance
- appearance
- sustainability.

Compressive strength

There is generally a good correlation between the density and strength of clay bricks. *BS EN 772-1:2011 Methods of test for masonry units.*

Table 7.1 Declared compressive strength of masonry units (N/mm²)

Masonry unit	Clay brick and block BS EN 771-1		Calcium silicate brick BS EN 771-2		Concrete brick & block BS EN 771-3	Autoclaved aerated concrete block BS EN 771-4
Condition A	Group 1	Group 2	Group 1	Group 2		
Brick	6.0	9.0	6.0	9.0	6.0	–
Block	5.0	8.0	–	–	2.9	2.9
Condition B						
Brick	9.0	13.0	9.0	13.0	9.0	–
Block	7.5	11.0	–	–	7.3	7.3
Condition C						
Brick	18.0	25.0	18.0	25.0	18.0	–
Block	15.0	21.0	–	–	7.3	7.3

Adapted from *BS EN 771-1 to -4*. Reproduced by permission of BSI.

Determination of compressive strength classifies bricks by their compressive strength into two groups: Group 1 units having not more than 25 per cent formed voids (20 per cent for frogged bricks) and Group 2 having formed voids greater than 25 per cent, but not more than 50 per cent. Furthermore the classification considers three conditions for bricks and blocks. Condition A accounts for masonry units in both leaves of a cavity wall for a single- or 2-storey building and the inner leaf for the second and third storeys in a 3-storey building. Condition B accounts for the external leaf in a 3-storey building and the internal walls in the ground storey of a 3-storey building. Condition C accounts for the inner leaf of a cavity wall in the ground storey of a 3-storey building (see Table 7.1).

Self-assessment question 7.4

Why should bricks in Group 2 have a higher compressive strength than bricks in Group 1 and why should bricks used in Condition B and Condition C have a higher compressive strength than bricks used in Condition A?

Flintlimes are generally stronger than sandlimes, and strength generally increases with age.

A further strength consideration for concrete bricks is covered by *BS EN 772-6:2001 Methods of test for masonry units. Determination of bending tensile strength of aggregate concrete masonry units*.

Water absorption

In clay bricks, the water absorption is dependent on the type of brick and the method of manufacture. *BS EN 772-21:2011 Methods of test for masonry units. Determination of water absorption of clay and calcium silicate masonry units by cold water absorption* considers the water absorption of clay and calcium silicate bricks. Most clay common and facing bricks have water absorption of between 12–26 per cent, whilst calcium silicate bricks have typical water absorptions of between 11–21 per cent dependent on the size and distribution of the pores (*BS EN 772-11 Methods of test for masonry units. Determination of water absorption of aggregate concrete, autoclaved aerated concrete, manufactured stone and natural stone masonry units due to capillary action and the initial rate of water absorption of clay masonry units*). Concrete bricks have similar absorption rates to those of calcium silicate bricks (*BS EN 772-7:1998 Methods of test for masonry units. Determination of water absorption of clay masonry damp proof course units by boiling in water*). The water absorption should be a maximum of 4.5 per cent for Class A engineering bricks, or bricks used for damp proof courses, and a maximum of 7.0 per cent for Class B engineering bricks.

Durability

Durability is affected by the water absorption of the brick, but in clay bricks a more important consideration is that of the soluble salt content of the brick, which could cause staining or *efflorescence* when the salts are dissolved by rainwater absorbed by the brick, which are then deposited on the surface of the brick when the rainwater evaporates. Soluble sulphate salts may attack the cement in mortars, causing the mortar to expand. Both of these issues will be considered in more detail later in this chapter.

The degree of burning affects the chemical and frost resistance of clay bricks. Under-burnt bricks are less durable than well-burnt bricks. *BS EN 771-1* specifies three levels of frost resistance for clay bricks. F0 designates bricks used in passive exposure conditions, that is generally for internal use and the internal leaf of cavity walls. These bricks should not be used externally unless protected by impermeable cladding or render. F1 designates bricks used in moderate exposure conditions, that is where bricks are durable apart from when the bricks are in a saturated condition or subject to repeated freezing and thawing. F2 designates bricks used in severe exposure conditions, that is where they are considered durable in all situations. Engineering bricks have the best durability of clay bricks in very exposed conditions. Bricks designated as having low soluble salt content should not contain more than 0.03 per cent by weight of soluble magnesium, potassium and sodium salts, 0.3 per cent by weight of soluble calcium salts and not more than 0.5 per cent by weight of soluble sulphate salts.

Frost damage to bricks causes the face to break away or *spall* as the ice crystals expand and leaves the inside of the brick exposed to further frost damage and water penetration.

Calcium silicate and concrete bricks do not suffer from soluble salt contamination in the same way that clay bricks do. Repeated wetting, drying and freezing also have little effect on either of these types of bricks. This is considered in *BS EN 772-18:2011 Methods of test for masonry units. Determination of freeze–thaw resistance of calcium silicate masonry units.* However, although calcium silicate bricks do not contain any harmful salts they can be susceptible to attack from sulphur dioxide present in the atmosphere or contact from harmful salts from soils or industrial waste.

Thermal and moisture movement

Moisture movement is negligible in well-burnt clay bricks. However thermal movement can be as much as 0.3mm per metre run of wall. It is therefore advisable to incorporate 10mm wide expansion joints containing compressible material, protected by a mastic sealant, for every 12m of wall not broken up by door or window openings. In house construction this figure is rarely reached. This is considered in *BS EN 772-19:2000 Methods of test for masonry units. Determination of moisture expansion of large horizontally perforated clay masonry units*.

The moisture movement of calcium silicate and concrete bricks is greater than that for clay bricks. In consequence it is recommended that 10mm expansion joints be provided for every 7m of unbroken calcium silicate or concrete brickwork. This is considered in *BS EN 772-10:1999 Methods of test for masonry units. Determination of moisture content of calcium silicate and autoclaved aerated concrete units*. Thermal movement is similar to that for clay bricks. It is inadvisable to bond clay, calcium silicate or concrete bricks together, due to their differing movement coefficients.

Thermal conductivity

Clay bricks generally have a high thermal conductivity and thus a poor thermal insulation value. The thermal conductivity value varies with the type and density of brick, its moisture content and whether it is in a protected or unprotected position between 0.36W/mK and 0.96W/mK.

The thermal conductivity of calcium silicate varies between 1.04W/mK and 2.06W/mK, dependent on the density and position of the brick.

Sound insulation

The sound insulation value of brickwork is directly proportional to the density of the wall. Most heavy clay, calcium silicate or concrete bricks can provide good sound reduction values in the range of 44 to 50dB.

Fire resistance

The fire resistance of clay bricks is generally good, since the bricks have been through the refractory process in a kiln and fired at a higher temperature than that which normally would occur in house fires.

Similarly calcium silicate bricks are capable of providing fire resistance up to 3 hours in cavity walls or 1B solid walls. For higher fire resistance requirements the thickness of the wall would need to be increased.

Concrete bricks have high fire resistance properties too.

Appearance

Applied finishes and textures are available for clay facing bricks. Sand is often used for texturing and can assist in releasing handmade bricks from their timber moulds. Colouring may be achieved either by natural pigments contained in the clay or by the addition of oxide pigments.

Calcium silicate bricks are generally off-white or pale-pink in colour, although other pastel colours may be obtained by the use of pigments. The bricks generally have a smooth finish, but textured finishes are available.

Concrete bricks tend to have a slightly roughened surface texture and they are available in smooth, rustic or weathered finishes. Different colours are produced by the addition of pigments into the mix.

Sustainability

The extraction process of the raw material for clay bricks uses energy, and quarrying can destroy natural habitats and deface the landscape. However, disused quarries have been turned into water parks or used as landfill sites for the disposal of refuse.

Most brick factories are situated close to the quarries, so energy costs in transportation of the raw materials to the factory are low.

Bricks have a high amount of embodied energy through the heat required first in the drying process and then later in the firing process. The production of this heat involves the burning of fossil fuels and the emission of both CO_2 and other pollutants to the atmosphere. However brick manufacturers, like cement manufacturers, have reduced their dependence on fossil fuels and are now using alternative fuels and finding ways of making fuel savings.

Bricks are durable and can be reclaimed from demolition sites and re-used or crushed and recycled to provide hardcore or aggregates for concrete with high fire resistance properties. Furthermore, most bricks are made in the UK and very few are imported, so the energy expended in transportation is lower than it is for imported materials. However the density of bricks means that energy expended in transportation from the point of manufacture to the point of use can be high.

With calcium silicate bricks, although energy is expended in the autoclaving process it is less than that used to manufacture clay bricks. Also, less CO_2 and other pollutants are emitted into the atmosphere. In consequence calcium silicate bricks are generally considered to have a lower environmental impact than their clay counterparts.

The sustainability credentials for concrete bricks are similar to those for concrete.

Efflorescence

When brickwork becomes wet, the porous bricks absorb some of the water. This water can dissolve some of the soluble salts contained in clay bricks. These soluble salts are impurities that are contained within the clay from which the bricks have been made. As weather conditions change, solar radiation and drying winds can cause evaporation of moisture from the external surface of the bricks. As the moisture in the brick is brought to the surface by this process the dissolved salts are brought to the surface too. When the water evaporates from the surface the salts are left behind and crystallise as a white deposit. If the crystallisation process occurs on the surface of the bricks it merely looks unsightly, but if the crystallisation process occurs immediately beneath the surface then spalling may occur, similar to that caused by frost damage.

It is difficult to cure efflorescence, but it should eventually disappear when all the soluble salts in the brickwork have been brought to the surface. This may take many years. Specifying bricks having a low soluble salt content or using calcium silicate or concrete bricks that do not contain soluble salts can significantly prevent the problem.

Sulphate attack

Particularly aggressive soluble salts that can be found in clay bricks are the sulphate salts of calcium, sodium, magnesium and potassium. When these dissolve in water absorbed into the bricks, they may migrate and attack a component of Portland cement used in mortar called *tricalcium aluminate*. This causes the mortar to expand and eventually to soften and crumble. This expansion of the mortar can result in a 2 per cent vertical and 1 per cent horizontal expansion in the brickwork that can cause stresses to occur, particularly around openings where the external leaf of the cavity wall may expand, but the internal leaf, which is unaffected by the sulphate attack, remains static. The expansion of the external brickwork may also put pressure on roof components at the top of a wall.

Self-assessment question 7.5

Why should the internal leaf of a cavity wall be unaffected by sulphate attack?

The problems of sulphate attack may be lessened by selecting bricks with a low soluble salt content, although sometimes the sulphate salts may not have been contained in the bricks but may have come from the soil that the brickwork is in contact with or the sand used in the mortar. Another good measure for reducing the incidence of sulphate attack is to use Sulphate Resisting Portland Cement (SRPC) in the mortar. This mortar contains less tricalcium aluminate, the component in cement that the sulphate salts react with, than Ordinary Portland Cement (OPC). Yet again calcium silicate and concrete bricks do not contain soluble sulphate salts.

A good design measure to combat the incidence of efflorescence and sulphate attack in brickwork is to use overhanging eaves, verges and cills wherever possible to shelter and protect the brickwork beneath from the effects of rain penetration.

Blocks

Blocks are defined as units exceeding the work sizes of a brick, with a minimum height of 190mm, but finer definitions are given in the British Standard Specification associated with each type of block. Clay blocks are covered by *BS EN 771-1:2003 Specification for masonry units. Clay masonry units*. Concrete blocks are covered by *BS EN 771-3:2011. Specification for masonry units. Aggregate concrete masonry units (dense and lightweight aggregates)* and *BS EN 771-4:2011 Specification for masonry units. Autoclaved aerated concrete masonry units*.

Blocks have six unique characteristics:

- They are light in weight, enabling their greater bulk to be handled easily when constructing a wall.
- They are sized to correspond to multiple bricks, enabling bonding in and work requiring corresponding course levels to be undertaken.
- They are comparatively cheap to produce and lay, giving economical wall construction in comparison with brickwork.
- They generally have better thermal resistance values than bricks.
- They accept nail or screw fixings more readily than brickwork.
- They have a keyed or textured surface for the application of plastered or rendered finishes.

Self-assessment question 7.6

Where might work requiring corresponding course levels between blockwork and brickwork be required?

Types of blocks

Blocks may be divided into two main materials:

- clay
- concrete.

Clay blocks

These are manufactured in a similar manner to clay bricks, but are generally extruded hollow units of honeycomb construction with keyed surfaces for the application of internal plaster or external render (see Figure 7.2). They are available in three sizes, 290mm long x 215mm high and 62.5, 75, 100 or 150mm wide. The 62.5 and 75mm thicknesses are only suitable for non-loadbearing applications, such as internal partition walls. These will be considered in more detail in Chapter 20 (internal walls and partitions).

Clay blocks are particularly brittle and difficulty can be experienced in cutting them to the required size. For this reason, special half and three-quarter blocks are produced for the purposes of bonding.

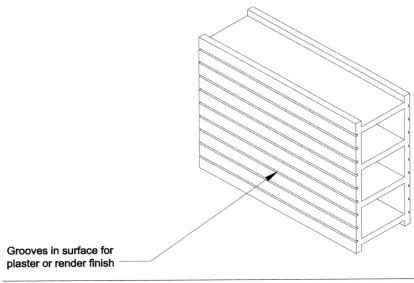

Grooves in surface for plaster or render finish

Figure 7.2 Typical clay block

These blocks are rarely used these days in the United Kingdom, but they are still used regularly in other parts of Europe for the construction of both internal and external walls.

Concrete blocks

These may be produced from either dense or lightweight concrete. The lightweight blocks are particularly useful for the inner leaves of cavity walls and internal partition walls and may be manufactured from lightweight aggregates, with or without *fines* (aggregates less than 5mm in size), or alternatively can be produced from autoclaved aerated concrete.

BS EN 771-3:2011 defines concrete blocks as a walling unit of length, width or height greater than that specified for a brick. Its length should not exceed 6 times its thickness and no dimension should be greater than 650mm. Blocks are classified as solid, cellular or hollow:

- *solid* – containing no formed holes or cavities
- *cellular* – containing one or more formed holes or cavities that do not completely pass through the block; the closed end is the upper surface on which the mortar is laid.
- *hollow* – containing one or more formed holes that pass completely through the unit (see Figure 7.3).

There are a wide range of sizes available from 390–590mm in length, 140–290mm in height and 60–250mm in thickness. Blocks having a thickness greater than 75mm are suitable for loadbearing purposes. Blocks having a minimum density of 1500kg/m³ are suitable for general use in building, including use in the ground below the damp proof course. However, other concrete blocks may be used below ground level if the manufacturer can provide evidence that the blocks are suitable for the purpose that they are intended to be used.

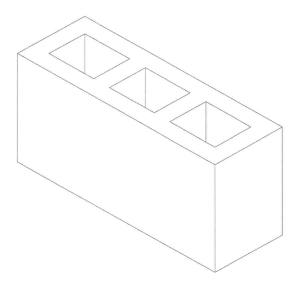

Figure 7.3 Typical hollow concrete block

Manufacture of concrete blocks

For dense and lightweight aggregate blocks the mixture of cement, the appropriate aggregates, either dense or lightweight, and water is placed in a steel mould and compacted by pressure or vibration. The blocks are de-moulded immediately after compaction and then left to *cure*, or set, either naturally or artificially under high temperature and humidity.

Autoclaved aerated concrete blocks are made from a mixture of cement, sand or *pulverised fuel ash* (PFA) (see Figure 7.4) admixtures to aerate the mix and water. The PFA is first mixed with sand and water to form a slurry. This is then heated before being mixed with cement, lime

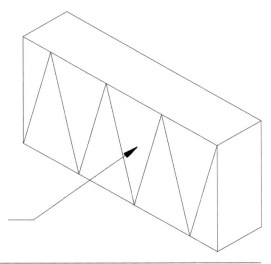

Scratched surface to receive
plaster or render finish

Figure 7.4 Aerated concrete block

and a small amount of powdered aluminium sulphate. The mixture is poured into a large steel mould and the aluminium sulphate reacts with the lime to produce hydrogen gas which is given off to be replaced by air which forms bubbles within the block. Following the initial partial set, the concrete slab that has been produced is cut into blocks by means of taut wires. The blocks are then autoclaved in a similar manner to calcium silicate and concrete bricks to develop strength.

These blocks are extremely light in weight, having a density as low as 475kg/m^3, and they have high thermal resistance properties and can be easily cut with the aid of a specially designed saw. However, they have high water absorption characteristics and must therefore be stored carefully prior to laying.

Properties of blocks

The following physical characteristics will be considered for each type of block:

- compressive strength
- water absorption
- durability
- thermal and moisture movement
- thermal conductivity
- sound insulation
- fire resistance
- appearance
- sustainability.

Compressive strength

BS EN 771-1:2003 specifies a minimum average compressive strength for clay blocks of 1.4N/mm^2 for non-loadbearing applications and 2.8N/mm^2 for loadbearing applications.

BS EN 771-3:2011 specifies compressive strength classifications for concrete blocks over 75mm in thickness. These are 2.8, 3.5, 5.0, 7.0, 10.0, 15.0, 20.0 and 35.0N/mm^2 and lightweight aggregate blocks have strengths up to approximately 10.5N/mm^2. Aerated concrete blocks manufactured to *BS EN 771-4:2011* have strengths up to approximately 5.5N/mm^2.

Water absorption

Blocks, because of their more porous composition and because they generally contain holes or cavities in their construction, have higher water absorption characteristics than bricks. This is covered by *BS EN 772-11:2011 Methods of test for masonry units. Determination of water absorption of aggregate concrete, autoclaved aerated concrete, manufactured stone and natural stone masonry units due to capillary action and the initial rate of water absorption*. Concrete blocks using dense aggregates are better for use below ground level, since blocks containing lightweight aggregates are more porous. Aerated concrete blocks are particularly porous and care must be taken to protect them from water penetration prior to their installation.

Blocks may be used in external solid walls or the external leaf of a cavity wall, but they must be suitably protected from rain penetration by applying a suitable external finish such as cement:sand render, tile hanging or weatherboarding. This will be considered further in Chapter 22 (external wall finishes).

Durability

The durability of clay blocks is similar to that of clay bricks. However they have fewer problems regarding frost damage, efflorescence and sulphate attack, as they will be kept drier than their brick counterparts.

Concrete blocks should not contain soluble salts but the mortar used to join them may be subjected to sulphate attack in applications below ground from soluble sulphate salts that may be present in the soil. Below ground they may also be more susceptible to frost damage due to their relatively high moisture absorption characteristic.

Thermal and moisture movement

Thermal movements are negligible and similar to those for clay and concrete bricks. Moisture movement, however, is more problematic, particularly with aerated concrete blocks, and can lead to cracking of the blocks due to drying shrinkage if they have been allowed to become particularly damp prior to laying.

BS EN 772-14:2002 Methods of test for masonry units. Determination of moisture movement of aggregate concrete and manufactured stone masonry units and *BS EN 772-10:1999 Methods of test for masonry units. Determination of moisture content of calcium silicate and autoclaved aerated concrete units* stipulate that drying shrinkage should be a maximum of 0.09 per cent for aerated concrete blocks and 0.06 per cent for all other concrete blocks.

Thermal conductivity

Because of the porous composition of the blocks and the use of cellular or hollow construction in many of the blocks, they have relatively good thermal resistance. Indeed some cellular and hollow blocks have their cavities filled with extruded plastic materials in order to further improve their thermal resistivity. Some solid blocks have a 60mm layer of polystyrene bonded to their inner face.

Thermal conductivity is closely related to the material of manufacture, the composition and thickness of the block as well as the addition of further thermal insulation materials. Thermal conductivities for concrete blocks range from 0.70W/mK–1.28W/mK for dense cellular blocks, 0.11W/mK–0.20W/mK for lightweight cellular blocks and autoclaved aerated concrete blocks and 0.10W/mK for clay blocks.

Because of their low thermal values blocks are particularly suitable for use as the inner leaf of cavity walls.

Sound insulation

Clay and concrete blocks have lower sound insulation properties than bricks because of their lower density. The average sound reduction in decibels for a concrete block wall ranges from 42–49dB.

Fire resistance

The fire resistance of clay blocks is less than for the equivalent thickness of brickwork due to their cellular structure. A 100mm wide block may be considered to have a fire resistance of 2 hours.

A 100mm thick dense concrete block wall may be classified as having a 2-hour fire resistance if it is loadbearing and a 4-hour fire resistance if it is non-loadbearing. These values are enhanced if plaster is applied to the internal face of the block.

Appearance

Clay blocks have grooved surfaces to provide a good key for plaster or render. Dense and lightweight aggregate concrete blocks tend to have a slightly rough, open textured surface, which provides a good bond for applied finishes. Aerated concrete blocks have a smooth surface and consequently are generally scored to provide a key for applied finishes.

Concrete blocks may be supplied with exposed aggregate finishes for external walling.

Sustainability

Clay blocks have similar sustainability credentials to those of clay bricks, although because they are mostly imported the energy expended in transportation is higher, even though their density is less than that of bricks.

Concrete blocks have similar sustainability credentials to concrete, although dense aggregate blocks often use recycled aggregates and lightweight concrete blocks use lightweight aggregates derived from other processes such as fly ash and foamed granulated blastfurnace slag. The low density of lightweight aggregate concrete blocks and autoclaved aerated concrete blocks creates savings in energy expended through transportation.

Visit the companion website to test your understanding of Chapter 7 with a multiple choice questionnaire, and to see the following feature illustrated with a full-colour photo:

• **7.1** Medium density concrete block

CHAPTER 8

External walls

Introduction

The primary purpose of external walls is to enclose and protect the building. In domestic construction these walls may also be used to carry the loads from the upper floors and roof to the foundations.

In this chapter it is intended to consider the main functional requirements of external walls and to consider how walls have been designed to meet these functional requirements. Both solid and cavity wall construction will be considered. In designing the walls to fulfil the functional requirements there are occasions when the needs of one requirement may be in conflict with those of another. In these cases compromises have to be reached.

Functional requirements

The following functional requirements apply to the design of external walls:

- strength and stability
- weather resistance
- sound insulation
- thermal insulation
- fire resistance
- durability
- appearance.

Strength and stability

Because external walls in domestic construction are used to carry the loads from the upper floors and roof to the foundations, the walls need to be strong enough to resist the stresses applied by the dead, live and superimposed loads being carried.

The strength of the wall is determined by its thickness and also the compressive strength of the materials used in its construction. The thickness of the wall needs to be sufficient to keep the stresses imposed by the applied loads within the safe compressive strength of the constructional materials, and the mortar used to bond the materials together should be slightly weaker than the brick or block, so that if movement in the wall occurs due to settlement, the break will be concentrated within the mortar and not the brick/block. This allows repair to be undertaken more easily.

Bonding of bricks, both along the length of the wall and also through its thickness, assists the distribution of the applied load so that over-stressing of materials at specific points is avoided. Bonding is discussed in more detail in Chapter 9 (bonding and openings in walls).

The wall also needs to be stable enough to resist overturning due to the application of lateral forces or buckling due to the excessive slenderness of the wall.

Ideally the centre of gravity of any load carried by a wall should coincide with the centroid of area of the wall. However, this is not always possible. For instance where an external wall is supporting a timber floor joist carrying the load from an upper floor, the ideal location for the joist would be to sit over the entire thickness of the wall so that the centre of the bearing where the load is being transferred would coincide with the centre of the wall (see Figure 8.1). Unfortunately this is not practicable, since the end of the joist would be exposed to the external climate and the timber would be susceptible to wet rot.

To overcome this problem the end of the joist should be protected from the external climate by at least half the thickness of the wall. This will shift the centre of the bearing and thus the centre of gravity of the load to the inner quarter of the wall and away from its centroid of area (see Figure 8.2).

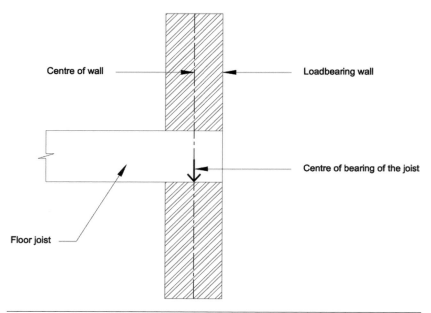

Centre of wall

Loadbearing wall

Centre of bearing of the joist

Floor joist

Figure 8.1 The centre of bearing of a floor joist coinciding with the centre of a wall

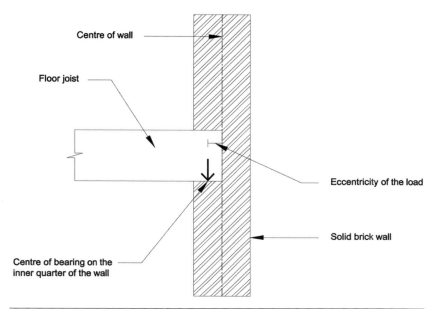

Figure 8.2 The centre of bearing of a floor joist in the inner quarter of a solid wall

The problem becomes more acute in cavity wall construction. Here the floor joist can only bear onto the inner leaf of the wall, as the central cavity has no loadbearing capabilities. Thus the distance between the centre of the bearing and the centroid of area of the wall (now in the centre of the cavity) is even greater (see Figure 8.3).

The most acute case is experienced when joist hangers are used to support a floor joist and transfer its load onto a cavity wall. The centre of gravity of the floor load here is again in the centre of the bearing onto the

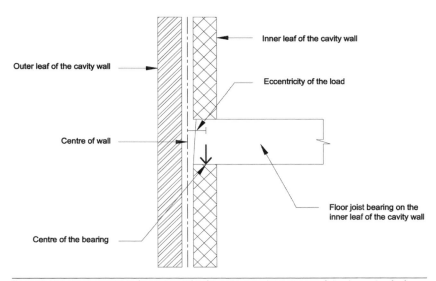

Figure 8.3 The centre of bearing of a floor joist in the centre of the inner leaf of a cavity wall

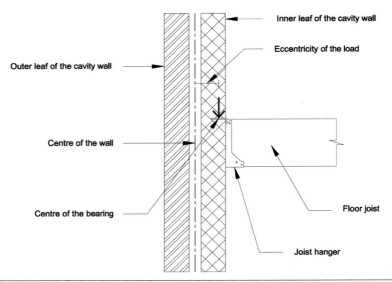

Figure 8.4 The centre of bearing of a floor joist on a cavity wall where joist hangers are used

wall, but in this case this point is even further away from the centroid of area of the wall than it was in the previous case (see Figure 8.4).

This disparity between the position of the centre of gravity of the load and the centroid of area of the wall is called the eccentricity of the loading. Clearly the case illustrated in Figure 8.4 displays the greatest degree of eccentricity.

Stresses set up by eccentric loading on walls are higher on the loaded side of the wall and lower on the unloaded side. This stress distribution presents a tendency for the wall to overturn. This has to be resisted.

Self-assessment question 8.1

What would be the eccentricity (the distance between the centre of gravity of the load and the centroid of area of the wall) of a cavity wall having a 102mm thick brick external leaf, a 75mm cavity containing a 50mm thick layer of insulation and 125mm thick concrete block internal leaf, with the floor joists bearing onto the block inner leaf of the wall?

If a wall is built excessively tall in relation to its thickness there will be a tendency for it to buckle under load. To resist this tendency the proportions of the effective height of the wall in relation to its effective thickness needs to be kept within acceptable limits. This is called the *slenderness ratio*.

$$\text{Slenderness ratio} = \frac{\text{Effective height of wall}}{\text{Effective thickness of wall}}$$

The effective height is taken as the distance between points of support or stiffening of the wall. In domestic construction this is normally the storey height of the building. The effective thickness of the wall is taken as the thickness of the structural materials and ignores the thickness of internal plasters and external rendering where these are used. It also ignores the thickness of the cavity. However, in a cavity wall, the external leaf may take some of the load that is imposed on the inner leaf of the wall by joining the two leaves together with wall ties. This will be considered further in this chapter when the construction of cavity walls is discussed. The effective thickness of a cavity wall for the purposes of calculating its slenderness ratio is therefore taken as the combined thickness of the two leaves plus 10mm width (Building Regulations Approved Document A, Section 2C8).

The maximum slenderness ratio for a wall should be 16.

Self-assessment question 8.2

What is the slenderness ratio of a cavity wall of 2.5m storey height and having a 23mm thick external rendering, 102mm thick brick external leaf, a 75mm cavity containing a 50mm thick layer of insulation, a 125mm thick concrete block internal leaf and a 13mm thick internal plaster finish?

Is it acceptable?

The combined dead and imposed load on the wall should not exceed 70kN/m at its base and the walls should not be subjected to lateral loading (apart from wind loading).

The stable height of an external wall will also be dependent upon factors such as the average wind speed at the site, the topography of the site (whether the building is built on a hill or near a ridge) and the altitude of the site. Other factors will be whether the site is in a rural or urban location and the distance that it is situated from the coast. It is recommended that storey heights should be a maximum of 2.7m and that the maximum height of a house, measured from the lowest finished ground level adjoining the building to the highest point of any wall or roof, should be 15m. In addition the height of the house should not greater than twice its width.

Cavity walls

These are constructed of two leaves of brickwork or blockwork not less than 90mm each in width, separated by an air space or cavity width (Building Regulations Approved Document A, Section 2C8). The air space must be a minimum of 50mm in width, but may be up to 150mm in width. In order to provide stability to the cavity wall, wall ties should tie the external and internal leaves together. These may be made from galvanised steel, twisted wire or stainless steel and spaced at 900mm horizontal and 450mm vertical maximum spacings (equivalent to 2.5 ties/m²) and it is recommended that this spacing be decreased for cavity widths between

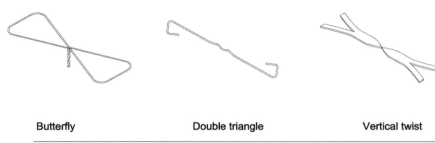

| Butterfly | Double triangle | Vertical twist |

Figure 8.5 Types of wall ties

75–150mm. The spacing of wall ties also needs to be decreased around openings and this is discussed in more detail in Chapter 9 (bonding and openings in walls). The length embedded in the masonry at each end of the ties should be a minimum of 50mm in both leaves of the cavity wall. There have been problems with some wall ties made from twisted wire or galvanised steel strip corroding over a period of time, thus wall ties made from stainless steel are to be preferred for durability. Metal ties conduct heat readily and may be a source of thermal bridging between the interior and exterior of the wall. For this reason polypropylene wall ties were developed, but they were found to be not as strong as metal ties, and have subsequently fallen out of use in recent years. The ties are generally of three distinct types specified in *BS 1243:1978 Specification for metal ties for cavity wall construction: butterfly type, double triangle type and vertical twist type*. Although *BS 1243* has now been withdrawn and the section on wall ties is now included in *BS EN 845-1:2003 + Amendment 1:2008 Specification for ancillary components for masonry. Ties, tension straps, hangers and brackets*, it is still referred to in the Building Regulations Approved Document A and the three types of wall tie to which it refers are still recognised and manufactured (see Figure 8.5). The butterfly type wall ties are really only suitable for cavity walls with a cavity width between 50mm–75mm. The double triangle type wall ties are suitable for cavity walls with a cavity width up to 100mm, whereas the vertical twist type, being the sturdiest, are suitable for all cavity widths. Both the butterfly and vertical twist type wall ties have the disadvantage of being difficult to use when partial fill rigid foam insulation boards are used, as there is no facility on these ties for a board retaining clip to prevent the board falling into the cavity. This is discussed further in the later section on thermal insulation.

Self-assessment question 8.3

Why should the maximum horizontal spacing between wall ties decrease when a 75mm–100mm wide cavity is adopted?

Although bricks may be used in both leaves of a cavity wall, lightweight concrete blocks are now commonly used for the construction of the inner leaf instead of bricks. This is because they can be laid more quickly than bricks (1 block is equivalent in size to 6 bricks) and they have a better thermal resistivity value than bricks and so make a better contribution to the overall thermal performance of the external wall. However there are disadvantages to the use of these blocks. In particular they have a lower compressive strength than bricks and this is important where floor joists are bearing their loads onto the inner leaf of the wall. Fortunately the loadings being applied in normal domestic construction are relatively light and within the compressive strength capabilities of these blocks. The manufacture and performance of these blocks was discussed in more detail in the previous chapter on bricks and blocks.

Below ground, the bricks in the outer leaf should be frost resistant and have a low soluble salt content. If blocks are to be used for the inner leaf of a cavity wall below ground they should have a density of $1500kg/m^3$ and a compressive strength not less than $7N/mm^2$. The cavity between the two leaves, if it is maintained below ground, can be filled with a weak mix concrete to prevent the two leaves being pushed together by the pressure of the surrounding ground (see Figure 8.6). This concrete should be terminated at least 225mm below the DPC to prevent rainwater from passing across the cavity and into the inner leaf of the wall.

Alternatively a solid wall constructed of brickwork or blockwork can be built from the foundations to within 75mm of ground level (see Figure 8.7). Where deep strip or trench fill foundations are used the top of the concrete foundation will come to this level anyway.

Weather resistance

Dampness may penetrate through the external walls into the house in three main ways:

- rain penetrating horizontally through the fabric of the wall
- the capillary rise of ground water
- rain penetrating vertically down from the head of the wall.

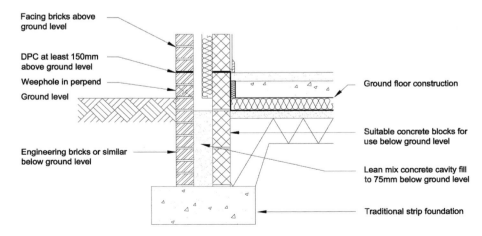

Figure 8.6 The base of a cavity wall where the cavity is maintained below ground level

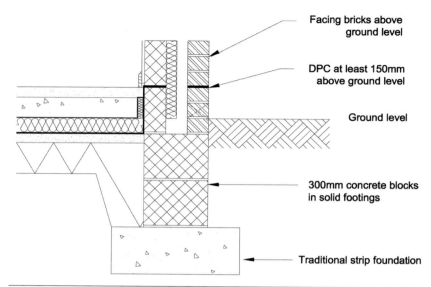

Facing bricks above
ground level

DPC at least 150mm
above ground level

Ground level

300mm concrete blocks
in solid footings

Traditional strip foundation

Figure 8.7 The base of a cavity wall with solid footings

Resistance to horizontal rain penetration

As discussed in Chapter 1 (introduction), the horizontal penetration of rain through the fabric of the wall may be resisted in one of three ways:

- By building the wall thick enough to prevent complete penetration. This relies on the weather conditions changing before the rain penetrating through the wall has the opportunity to reach the internal surface. Drying winds or solar heat will encourage evaporation of the moisture within the wall from its external surface, thus drying the wall out. If the wall is too thin then the rain penetrating through it may reach the internal surface before the weather change encouraging evaporation has had enough time to take effect. In the past, solid walls relied on this technique. However, severe rain penetration may occur through cracks in the walls caused by thermal movement during hot weather or subsidence caused by shrinkage of clay subsoils during drought.

- An impermeable barrier is provided to the external face of the wall so that rain is prevented from penetrating into the fabric of the wall. Applying *render* or *stucco* (a material used in the past to render walls and comprising a mixture of lime, sand and water, mixed with a small quantity of linseed oil to reduce its porosity) to the external surface of the wall could provide this external barrier. The main problem with this technique is that if the render cracks due to drying shrinkage, then rainwater can penetrate through these small hairline cracks and still enter the wall fabric and soak through to the internal surface. Unfortunately the process cannot easily be reversed by evaporation as the render acts as a barrier. Because of this the adoption of impermeable rendering has now lost favour and most modern renders are now permeable. This topic will be discussed in more detail in Chapter 22 (external wall finishes).

- An air cavity can be created between the moist external face of the wall and the dry internal face of the wall. Because rain cannot cross the air cavity then the internal surface of the wall will remain dry. The width of the cavity will be dependent upon the location of the building and its degree of exposure. This is the concept of the cavity wall. In sheltered areas the cavity need only be 50mm wide, but in more exposed locations the cavity will need to be considerably wider to prevent rain penetration from the exterior to the interior of the wall. Building Regulations Approved Document C, Diagram 12 divides the UK into zones of exposure and Table 4 provides guidance on the width of the cavity dependent upon the exposure zone in which the wall is located, the type of thermal insulation within the cavity and the type of facing to the external leaf of the cavity wall. Most modern houses built with loadbearing external masonry walls use this technique to fulfil the weather resistance requirement. For it to be fully effective there should be no path for the rainwater to cross the cavity. Unfortunately the cavity in the wall has to be closed where door and window openings are provided. This aspect will be considered further in the next chapter (bonding and openings to walls).

Additionally, where wall ties are used to strengthen the wall by binding together the external and internal leaves of masonry, there exists the opportunity for rainwater to cross the cavity via the wall ties. The wall ties must therefore be designed to have a drip at their centre to prevent them acting as a path for the rainwater across the cavity. It is also important when constructing the cavity walls to ensure that no mortar droppings are allowed to fall down the cavity and to lodge on the wall ties, otherwise this could provide an effective pathway for rainwater to cross the cavity by soaking through the dried mortar droppings. When cavity walls were first used it was believed that the cavity should be ventilated, so that moisture within the cavity itself could be dried. Thus air bricks were inserted at the top and bottom of cavity walls. This practice has now been abandoned as it was found that by ventilating the cavity the thermal performance of the wall was being detrimentally affected.

Self-assessment question 8.4

Why should ventilating the cavity have a detrimental effect on the thermal performance of a cavity wall?

Resistance to rising damp

Ground water may enter the wall at or near its base and creep up the wall by capillary action to enter the building above ground level. This rising damp could affect the wall for a height of up to 1m above ground level. The problem may be overcome by the use of a damp proof course or DPC. The DPC must be:

- impermeable to moisture
- continuous throughout the wall

- durable
- strong enough to support the loads carried by the wall without extruding
- unobtrusive when placed in position
- flexible enough to allow movement within the wall without being damaged
- comparatively simple to install.

There are eight different types of material in common use for damp proof courses and these may be divided into three broad categories:

- flexible
- semi-rigid
- rigid.

Flexible DPC materials

Lead – This is supplied in thin sheets to *BS EN 12588:2006 Lead and lead alloys. Rolled lead sheet for building purposes.* It is expensive, but will not extrude under normal loadings. Its durability can be reduced if it is placed in contact with fresh lime or Portland cement mortars. To reduce the effect of these mortars the lead sheet may be coated with bitumen paint.

Copper – This is similar to lead in that it is supplied in sheet form to *BS EN 1172:2011 Copper and copper alloys. Sheet and strip for building purposes*, and is expensive. It is however highly resistant to corrosion but needs protection from a bitumen paint coating where soluble salts are likely to be present in the masonry of the wall.

Bitumen felt – This is supplied in rolls and widths similar to the common widths of walls to *BS 6398:1983 Specification for bitumen damp proof courses for masonry*. The felt needs to be lapped by a minimum of 150mm at ends of rolls. It may extrude under heavy pressure from wall loadings but this is not normally a problem under light loadings generally found in domestic construction. Additional strength can be obtained by providing a lead, copper or aluminium core to the sheet. Bitumen felt is susceptible to damage and is brittle in cold weather and when exposed to prolonged amounts of ultra-violet light from the sun.

Polyethylene – This is similar to bitumen felt in that it is supplied in rolls to suit most wall widths to *BS 6515:1984 Specification for polyethylene damp proof courses for masonry*. However, it does not suffer from extrusion under normal loading conditions and there is no deterioration when it is placed in contact with other materials. It is now the most common material used for a DPC. Like bitumen felt it must be lapped a minimum of 150mm at ends of rolls.

Pitch polymer – A combination of coal tar pitch and a synthetic polymer base to *BS 743:1970 Specification for materials for damp proof courses*. It has similar characteristics to polyethylene but is far stronger and therefore less easily damaged.

Semi-rigid DPC materials

Mastic asphalt – This is a mixture of bitumen and inert mineral matter to *BS 6925:1988 Specification for mastic asphalt for building and civil engineering (limestone aggregate).* It is applied hot and spread to a thickness of 13mm in one coat. It does not deteriorate when placed in contact with other building materials but is likely to extrude when subjected to heavy loading.

Rigid DPC materials

Brick – Two courses of Staffordshire Blue engineering bricks having a maximum water absorption of 4.5 per cent by dry weight should be bedded in 1:3 Portland cement:sand mortar. The bricks are extremely strong and will not extrude under load or deteriorate when placed in contact with other building materials.

Slate – Laid in two courses in a similar manner to brick. The joints between slates should not be continuous through the two courses.

Polythene damp proof courses tend to be the most popular for house building.

The DPC should be installed in the wall at least 150mm above the level of the adjoining ground (Building Regulations Approved Document C, Section C5.5).

Self-assessment question 8.5

Why should the DPC be installed at least 150mm above the level of the adjoining ground?

Resistance to penetration downwards from the head of the wall

Most external walls to houses are protected at their head by the roof positioned on top of them. However, some external walls may be continued above the level of the edge of the roof to form a *parapet*. Where this occurs there is the probability that rainwater will soak down through the head of this parapet wall and cause dampness to occur in the building below. To prevent this, a DPC needs to be installed immediately beneath the *coping* that tops the parapet wall (see Figure 8.8).

Damp proof courses are also used around openings for the installation of doors and windows in cavity walls. This will be considered further in the next chapter (bonding and openings to walls).

In cavity walls, where rainwater has soaked through the external leaf of the wall, there needs to be provision at the foot of the cavity wall for this water to be able to drain from the cavity back to the exterior. This can be attained by providing *weepholes* at 900mm centres in the vertical joints between bricks at ground level (see Figure 8.6). Similar weepholes will also need to be provided above horizontal damp proof courses over the heads of openings in cavity walls.

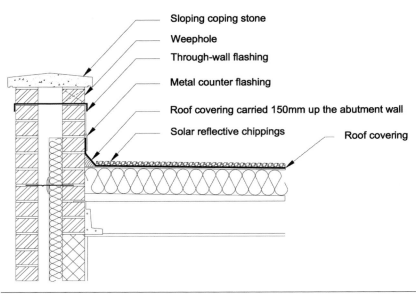

Figure 8.8 Weatherproof protection to a parapet wall

Sound insulation

The essential characteristics of sound transmission and insulation have previously been considered in Chapter 1 (introduction). In that chapter it was stated that sound is transmitted through a wall in two ways:

- Airborne sound – the sound waves impinge on the surface of the wall and cause the wall to vibrate like a diaphragm and so transfer the sound energy to the air on the other side of the wall.
- Structure borne or impact sound – the sound waves create vibration of the molecules within the structural materials of the wall. These vibrations can then be passed on to the molecules in other materials in contact with the wall. Thus impact sound can be passed from the external wall to the upper floor and roof structure and be transmitted throughout the building.

The two mechanisms of sound transmission require quite different treatments for insulation to reduce the amount of sound energy transferred. For airborne transmission the best method of insulation is to provide mass to the structure. The idea here is that the heavier the structure is, the less likely it is to vibrate as a diaphragm in sympathy with the sound waves impinging on its surface. For structure borne or impact sound transmission, the best mechanism for insulation is discontinuity or separation between the different parts of the structure, so that the sound energy cannot easily be passed from molecules in one material to the molecules in a neighbouring material. This can be difficult to adequately achieve, as the external walls in most houses fulfil a loadbearing function as well as an enclosing function, thus the upper floors and roof structures will be in direct contact with the wall for loadbearing purposes. The separation of the outer and inner leaf in a cavity wall does help to provide some discontinuity within the structure of the wall itself, but complete discontinuity between the two leaves of the wall is not possible since the wall ties that link the two leaves together are also likely to be able to provide paths for structure borne sound transmission across the cavity. In this respect the butterfly type and double triangle type wall ties will transmit less sound energy across the cavity than the sturdier vertical twist type wall ties.

Self-assessment question 8.6

Why should the use of the butterfly or double triangle wall ties in preference to the vertical twist type wall ties, for airborne sound insulation purposes, be a problem when considering the structural stability of the wall?

There are a number of other factors to be considered in the sound insulation of external walls. An external wall will need to prevent external noise (particularly traffic noise) from entering the house, but may also be required to prevent noise generated within the house from affecting the external environment. Furthermore, where properties are joined together as will be the case in semi-detached or terraced house construction, the separating or *party walls* between the properties will need to have a good standard of sound insulation to prevent the sound from one house being transmitted through these walls to the adjoining houses.

Another problem is that the insulation value of a material or an element of the building structure will vary according to the frequency of a sound. In other words some materials may insulate sound well at low frequencies, but not at higher frequencies and vice versa. Because the sensitivity to sound of the human ear is normally limited to frequencies within the range 100–3200 cycles per second, the performance of materials as sound insulators within this frequency range is considered to be most important. The unit used to measure the intensity of a sound is the decibel (dB). In considering requirements for sound insulation the intensity of sound that also takes account of the sensitivity of the human ear across different frequencies is used. This is referred to as the A weighting (dBA) and is used by the Building Regulations to determine the amount of sound insulation that should be provided by an external wall to reduce the transmission of external noise into the building. The performance standards for separating walls are given in the Building Regulations Approved Document E, Section 0 and sound insulation testing of the structure to ensure compliance with these standards is now required.

In Chapter 1 (introduction), it was stated that the level of sound insulation to an element would be reduced if there were gaps or cracks in the structure that would allow the easy transmission of sound waves through the structure. All joints should be filled with mortar, and frogged bricks should be laid with their frog uppermost to achieve the maximum mass per unit area for the wall. Where external walls are constructed from semi-porous materials, such as concrete blocks, then the sound insulation of the wall can be improved by sealing the surface of these blocks with a plaster coating. Where a separating wall joins a cavity wall the latter should be stopped with a flexible closer at the cavity, unless the thermal cavity insulation fully fills the cavity.

In addition, the total sound insulation of an external wall will only be as good as the weakest element of the wall. Doors and windows will have less mass than the masonry wall in which they are fixed, thus the total sound insulation of an external wall will be affected by the amount of door and window openings it contains.

Sound absorption is not the same as sound insulation. The purpose of sound absorption is to reduce reflection of sound from a wall surface; the purpose of sound insulation is to reduce the transmission of sound through the wall. Soft quilts may be added to the inner surface of a wall and this will significantly reduce the amount of sound energy reflected back into the room from that surface. Although these quilts can absorb up to 80 per cent of the sound energy impinging on the wall, the 20 per cent that will still pass through the wall can be sufficient to cause a noise nuisance.

Thermal insulation

The thermal insulation of a building element is measured by the air-to-air transmission coefficient (U-value). This is made up from the thermal resistivity values (the resistance to the passage of heat) of all the materials and air spaces that make up the building element being considered. The U-value of an external wall is also influenced by its degree of exposure, the amount of solar radiation it receives, the wind and moisture conditions of the climate (a damp material is less efficient as a thermal insulator than it is when it is dry). Because the U-value is a reciprocal value of the sum of all the thermal resistivity values of the various materials contained in the element, a lower U-value represents a higher degree of resistance to the passage of thermal energy.

The Building Regulations Approved Document L1A Section 4.21 sets out limiting U-values for various elements of the building. These are weighted average values, as explained in Chapter 1 (introduction), and for external walls this value is 0.30 W/m²K. In addition Building Regulations Approved Document C, Section 5.36 stipulates that the external wall should be designed and constructed such that its U-value does not exceed 0.7 W/m²K at any point in order to provide resistance to the development of surface condensation and mould growth.

Similarly to sound insulation, the overall thermal insulation of an external wall will be reduced where large numbers of openings containing doors and windows are contained within the wall, as these elements will have higher U-values than the masonry wall.

In cavity walls the air contained in the cavity is a good insulator, provided it is still, and will offer good resistance to heat flow through the wall. It follows that the wider this cavity is, the better will be the thermal insulation provided by the cavity. However, where cavities are ventilated to help rid them of the moist air that may be present when the outer leaf of the wall becomes saturated, the thermal insulation value of the cavity is likely to be adversely affected (see self-assessment question 8.4).

The use of lightweight concrete blocks for the construction of the inner leaf of a cavity wall can improve the thermal insulation value of the wall as the blocks have a higher thermal resistivity value than that of clay bricks and also concrete blocks containing dense aggregates.

Self-assessment question 8.7

Why is it that lightweight concrete blocks have a better thermal resistivity value than clay bricks have?

Further improvements to the thermal insulation value of an external wall can be made by adding extra thermal insulation material into the construction to:

- the external face of the wall
- the cavity, either by fully or partially filling it
- the internal face of the wall.

Thermal insulation materials will contain large amounts of air and are therefore particularly porous; they must therefore be protected from becoming damp through the absorption of large amounts of moisture. Typical materials may be expanded synthetic materials supplied as solid blocks, beads or injected foams, glass fibre quilts or mineral 'wool' supplied as quilts or blown fibres.

Thermal insulation applied to the external face of the wall

This is generally used on solid walls where no cavity exists to place the extra insulation. The solid blocks of expanded synthetic material, such as Polyisocyanurate, are secured to the external face of the wall and covered with a mesh that provides a good key for a thin coat of external rendering (see Figure 8.9). This coating serves a dual purpose; firstly it protects the insulation beneath from water penetration and secondly it provides an attractive surface finish to the wall.

Disadvantages of this technique are that the insulation is vulnerable to impact damage, and the external appearance of the house will be affected by the application of rendering, which may not be suitable in some areas.

Thermal insulation applied partially within the cavity

This technique is favoured on most new houses. Thermal insulation in the form of expanded plastics material, glass fibre or mineral wool *batts* (small blocks of material) is installed into the cavity as the wall is constructed. The batts are secured against the external surface of the internal leaf (facing the cavity) by means of plastic wheels fixed to specially designed wall ties. A gap of at least 25mm (as mentioned

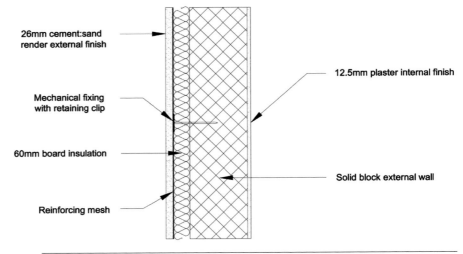

26mm cement:sand render external finish

Mechanical fixing with retaining clip

60mm board insulation

Reinforcing mesh

12.5mm plaster internal finish

Solid block external wall

Figure 8.9 Thermal insulation applied to the external face of the wall

previously) is maintained in the cavity between the outer face of the insulation material and the inner face of the external leaf (see Figure 8.10). This gap maintains the discontinuity between the two leaves of the cavity wall for weather resistance purposes. By widening the cavity a greater thickness of insulation material may be installed and a lower U-value can be obtained.

The disadvantages of this technique is that great care has to be taken during the installation of the insulation material within the cavity that it does not fall across the cavity and provide a bridge for the transfer of moisture. Additionally, this technique is only available with new construction and cannot be applied to previously constructed cavity walls, where the thermal insulation may need to be improved. Furthermore, it has been found that if the thermal insulation within the

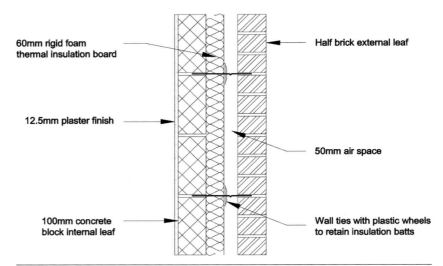

60mm rigid foam thermal insulation board

12.5mm plaster finish

100mm concrete block internal leaf

Half brick external leaf

50mm air space

Wall ties with plastic wheels to retain insulation batts

Figure 8.10 Partial cavity fill

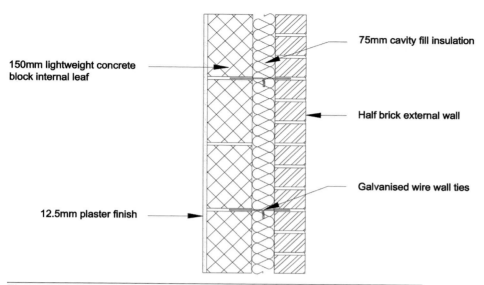

150mm lightweight concrete block internal leaf

75mm cavity fill insulation

Half brick external wall

Galvanised wire wall ties

12.5mm plaster finish

Figure 8.11 Full cavity fill

cavity is very efficient then the temperature of the external leaf of a cavity wall can become quite low, making the clay brickwork in the external leaf more susceptible to frost damage during periods of cold weather.

Thermal insulation applied fully within the cavity

This technique is used in 'retro-fit' situations; that is in situations where additional thermal insulation needs to be applied to a previously constructed cavity wall. Holes are drilled into the mortar joints of the external leaf of the wall at pre-determined intervals and either polyurethane or urea-formaldehyde foam is injected into the cavity or polystyrene beads or mineral wool fibres are blown into the cavity (see Figure 8.11). When the cavity has been filled the injection holes are made good.

The disadvantages with this technique are that with filling the cavity there is the possibility that water may penetrate from the outer leaf to the inner leaf of the wall. To prevent this from occurring the polyurethane foam contains resins that repel water. However, as the injected foam solidifies it shrinks, and this shrinkage may be impaired by the rough internal surfaces of the masonry inner and outer leaves of the wall. Where this happens small cracks may occur within the foam that may provide paths for the capillary movement of water across the cavity. With the dry installation of polystyrene beads or mineral wool fibres, these are randomly blown into the cavity and voids between the individual beads or fibres will be sufficiently large enough to prevent capillary paths from occurring. A resin is used to bind the randomly distributed polystyrene beads or mineral wool fibres together so that they will not cascade out of the cavity should a hole or opening be placed into the wall after installation.

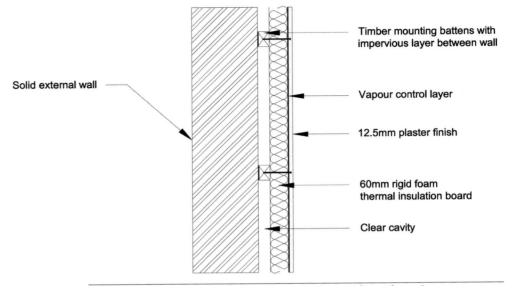

Solid external wall

Timber mounting battens with impervious layer between wall

Vapour control layer

12.5mm plaster finish

60mm rigid foam thermal insulation board

Clear cavity

Figure 8.12 Thermal insulation applied to the internal face of a wall

Thermal insulation applied to the internal face of the wall

Similar to the technique of applying insulation to the external face of the wall it is possible to apply the insulation to the internal face. Sheets of expanded plastics materials or glass fibre or mineral fibre quilts may be secured to the internal surface of the wall and then covered with a plaster or plasterboard finish (see Figure 8.12). Once again this is particularly suitable for solid walls where there is no opportunity of putting the extra thermal insulation within the cavity.

Disadvantages of this technique are mainly that it decreases the dimensions of the room and interstitial condensation may occur within the wall if a vapour barrier is not provided on the warm side of the insulation. This problem will be considered in more detail in Chapter 15 (timber frame construction). Many modern plasterboards now incorporate a layer of thermal insulation and a vapour control layer bonded to their outer surface, which is intended to face the external wall. These are considered in more detail in Chapter 21 (internal finishes).

Fire resistance

The main concern with regard to the fire resistance of an external wall is its ability to withstand the action of fire encroaching onto the property from outside or to withstand the spread of fire from one property to another and to provide sufficient stability to the floors and roofs it is supporting to allow fire fighting services to tackle the fire as effectively as possible. External walls should be constructed so the risk of ignition from a source outside the building, and the spread of fire over their external surfaces is restricted by their low rates of heat release in a fire.

Provided the external wall is situated more than 1m from any external boundary to the property then the minimum periods of fire resistance that should be exhibited by an external wall to a house as specified in the Building Regulations Approved Document B, Section 8.3 are 30 minutes where the top storey is not more than 5m above ground level and 60 minutes where the top storey is more than 5m above ground level. If the external wall is situated less than 1m from the boundary of the property, then the combustibility of the wall needs to be further restricted in order to reduce the susceptibility of its external surface to ignition from a source outside the building.

Masonry walls, either built from solid brickwork or of cavity wall construction with a brick outer leaf and lightweight concrete block inner leaf, have good fire resistance because they are manufactured from materials that have been through the refractory process (i.e. they have been produced in a kiln at high temperatures). The fire resistance of a wall generally increases in direct proportion to its thickness. A typical one brick thick wall or a cavity wall having a half brick external leaf, a 50mm cavity and a 100mm lightweight concrete block internal leaf can normally achieve 6-hours fire protection. This figure can be further improved if the internal surface of the wall is plastered, particularly if lightweight plaster is used.

The cavity within a cavity wall provides a route for the spread of smoke and flames, should a fire occur within the building. In order to restrict this spread, cavities need to be sub-divided and their edges closed at openings with cavity barriers (see Figure 8.13). These cavity barriers need to be firmly fixed so that their performance is not impaired by movement of the building or failure in a fire of their fixing. Often pre-formed cavity closers are used, which can double up as a barrier against thermal

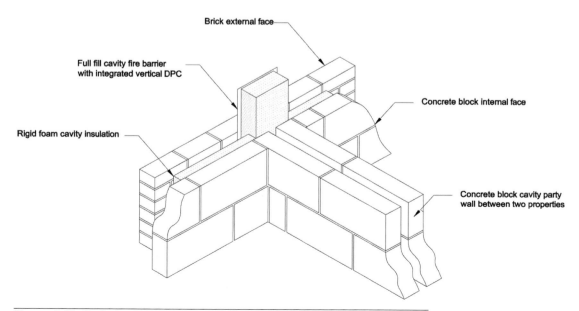

Figure 8.13 Cavity barrier to prevent the spread of fire within a cavity wall

bridging and also provide an effective barrier to the spread of fire within the cavity.

As with sound insulation and thermal insulation, the performance of the external wall with regard to fire resistance is reduced when windows and doors are included. The amount of windows in an external wall that is situated close to a boundary need to be restricted in order to limit the amount of thermal radiation that could pass through the window should a fire occur within the building.

Durability

The durability of an external wall is affected by three main weathering effects:

- frost action when clay bricks are saturated
- disruption by the crystallisation of soluble salts deposited on or just beneath the external surface of the brickwork by the evaporation of moisture, called efflorescence
- expansion of mortar caused by the chemical action of soluble sulphate salts diffusing into the mortar from the clay bricks when the brickwork remains wet for long periods.

All three of the above effects were considered in detail in the previous chapter on bricks and blocks.

Deterioration in all of the above cases is primarily caused by moisture entering the wall. This may be reduced considerably by providing overhangs at the eaves, lintels and sills to shelter the wall beneath from the effects of rain.

Appearance

There are a large number of different types of brick available, varying in colour and surface texture. Different coloured mortars are also available to complement and enhance the colour of the brickwork. There are also different styles of pointing used for the mortar joints (see Figure 8.14).

Patterns are provided by the different bonding arrangements of the bricks in the wall, particularly where the bricks have a slightly different colour on their header faces to their stretcher faces. Bonding will be considered in more detail in the next chapter.

All of these factors affect the appearance of brick external walls, but other external finishes are also used on walls such as rendering, tile hanging and weatherboarding. These techniques will be considered further in Chapter 22 (external wall finishes).

Appearance to external walls is also affected by the arrangement of windows (fenestration) and the use of precast concrete, stone or brick cills, copings, arches and lintels. These will be considered further in the next chapter (bonding and openings in walls).

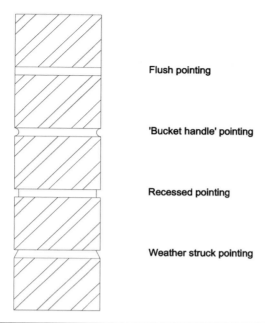

Flush pointing

'Bucket handle' pointing

Recessed pointing

Weather struck pointing

Figure 8.14 Styles of pointing

Visit the companion website to test your understanding of Chapter 8 with a multiple choice questionnaire, and to see the following features illustrated with full-colour photos:

- **8.1** Brick external skin with blockwork cavity
- **8.2** Partial fill cavity external wall
- **8.3** Partial filled cavity insulation board
- **8.4** Stainless steel wall tie system to connect new cavity wall to existing external wall

CHAPTER 9

Bonding and openings in walls

Introduction

Masonry walls can be built from bricks, blocks and stones. The most common method of external masonry wall construction uses bricks and blocks. This chapter will therefore concentrate mainly on wall construction using these materials. Walls built from masonry need to be joined together by a *mortar* that will allow the masonry units to be adjusted to line and level when they are being laid and will then stiffen and harden to enable the masonry units to work together within the wall. Overlapping the masonry units in a bond pattern in order to transfer the load effectively along and through the wall enhances this. There are a number of bonds that may be used for solid walls and cavity walls. This chapter concentrates on the five most common types of bond and also introduces special bricks that are either made to a particular shape or cut from whole bricks in order to maintain the bond pattern of the wall, particularly where the wall forms right-angled corners.

Where windows and doors are installed in external walls there is a need to form openings in the wall for these elements. The openings comprise a lintel or arch at the head of the opening, a reveal at the sides of the opening and a cill or threshold at the foot of the opening. The wall will require special treatment at these points in order to maintain its functional requirements, particularly strength and stability and weather resistance. This chapter reviews some of the common techniques used for both solid and cavity walls at the head, reveals and cill.

Brick bonding

To provide stability to a masonry wall the bricks, blocks or stones from which the wall is constructed must be fixed together by means of *mortar*, which is a mixture of cement, sand, lime and water. Mortars should fulfil the following requirements:

• Provide adequate strength, but not a greater strength than that required for the location or purpose of the wall.

- Provide good workability and plasticity to enable the bricks to be bedded and adjusted to the required line and level and then be able to stiffen and set in a reasonable time period.
- Provide adequate durability for the particular location and purpose of the wall, in particular provide good resistance to frost attack, efflorescence and sulphate attack.
- Provide a good bond with the bricks, blocks or stones being used to construct the wall.
- Prevent water penetration through the joint when hardened.
- Accommodate basic thermal and movement stresses without cracking.
- The materials used should be readily obtainable and economic.

The proportioning of the materials in the mortar will be dependent upon the loads and exposure that the wall may be subjected to. This may vary from 1:1:5–6 cement:lime:sand for normal masonry work where there is little chance of frost attack, to 1:3 cement:sand for walls where extra strength is required. The mortar should not lose the water from its mix easily when it is laid on dry bricks or blocks. Stiffer mortars, having less water in their mix, are recommended for masonry units having less absorption, such as stones. Strong mortars are not always the best to use in wall construction, as they tend to concentrate the effect of differential movement causing a few wide cracks to occur. Weaker mortars, however, will take up small movements so that any cracks are much smaller and less troublesome. The introduction of lime to mortar assists in providing good workability and plasticity to the mortar so that the masonry units can be bedded, and adjusted for line and level before the mortar sets. Materials called *plasticisers* can also be added to mortars to improve their workability and plasticity. These work by introducing air bubbles into the mortar, which break down the surface tension so that less water can be used in the mix to achieve the same level of workability. These air bubbles also help to give the mortar more frost resistance. Where there is a likelihood of sulphate attack from the soluble sulphate salts within the clay bricks of a wall, the mortar should contain sulphate resisting cement. This was discussed in Chapter 7 (bricks and blocks).

To spread the load evenly throughout the wall, the bricks must be laid to a regular pattern or bond so that the overlap of bricks in one course is achieved above and below in preceding and succeeding courses. This overlap should never be less than one quarter of a brick length and allows the load to be spread through a triangular distribution (see Figures 9.1–9.5).

The common bonds used in brickwork are:

- Stretcher bond
- English bond
- Flemish bond
- English garden wall bond
- Flemish garden wall bond.

The component parts of a standard brick were considered in Chapter 7 (bricks and blocks). In describing the five main types of bond listed above, reference will be made in particular to the header face and stretcher face of the brick. In addition it is important to understand the terminology used in describing the thickness of a brick wall. The standard size of a brick, 215mm in length, 102.5mm in width and 65mm in height enables two bricks laid end on with their header faces showing and using a 10mm mortar joint between them, to equal the length of one brick laid with its stretcher face showing. Thus a 215mm thick wall is described as a 1B wall (a wall of one brick thickness) and a 102.5mm thick wall is described as a ½B wall (half brick thickness). A horizontal row of bricks is described as a course and is generally 75mm in height (65mm for the brick height and 10mm for the bedding course of mortar).

Self-assessment question 9.1

What would be the thickness of a 1½B wall?

Stretcher bond

This consists of all stretcher faces in every course with a half brick overlap on each brick (see Figure 9.1). This bond cannot spread the load through the thickness of a one brick wall or thicker wall and is therefore mainly used in half brick walls of the half brick thick external leaf of a cavity wall.

English bond

This consists of alternate courses of headers and stretchers providing an overlap of one unit over another in the length of the wall equivalent to one quarter of the length of a brick. In a solid wall the bricks are bonded through the wall as well as along it. It is not really suitable for use in half brick walls. Where the wall forms a right-angled corner a brick has to be

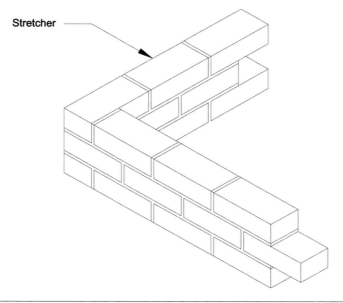

Stretcher

Figure 9.1 Stretcher bond to a half brick wall

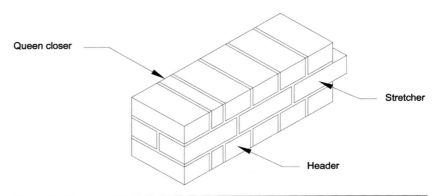

Queen closer

Stretcher

Header

Figure 9.2 English bond to a one brick wall

cut in half lengthways in order to maintain the quarter brick overlap for the bond. This brick is called a *queen closer* (see Figure 9.2).

Self-assessment question 9.2

Why should this bond not be suitable for use in half brick walls?

Flemish bond

This is similar to English bond except that each course consists of alternate headers and stretchers and is again mainly used in solid walls. It is not as strong as English bond but the appearance is often considered to be better. In order to form a stop end to the wall and still maintain the appearance of the bond both on the front face and the stop end face, two bricks need to be cut on a slanted or bevelled line so that one header end of the closer brick so formed is quarter brick thickness and the other is the full half brick thickness. This brick is called a *bevelled closer* (see Figure 9.3). In a 1½B wall a brick may need to be cut in half crossways to maintain the bond. This brick is called a *half bat*.

English garden wall bond

This is similar to English bond, but comprising three courses of stretchers to every one course of headers (see Figure 9.4). Sometimes, when the wall is only taking very light loads it may be built using five courses of stretchers to every one course of headers.

Flemish garden wall bond

This is similar to Flemish bond, but comprising one header to every three stretchers in each course (see Figure 9.5). Again, similar to English garden wall bond, if the wall is only taking very light loads it may be built using five stretchers to every one header in each course.

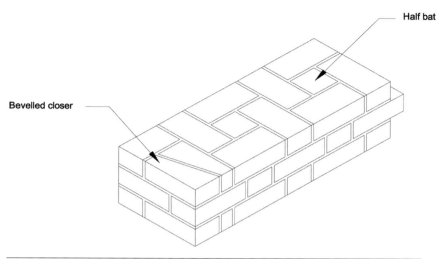

Half bat

Bevelled closer

Figure 9.3 Flemish bond to a one-and-a-half brick wall

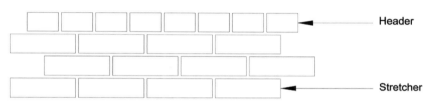

Header

Stretcher

Figure 9.4 English garden wall bond

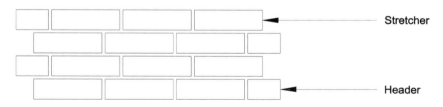

Stretcher

Header

Figure 9.5 Flemish garden wall bond

Openings in walls

Where doors and windows are to be located in walls, openings must be formed. The number, size and position of these openings must not impair the stability of the wall. Openings should not be greater in height than 2.1m and the sum of the width of openings in one wall should not exceed two-thirds of the unsupported length of the wall. In addition, openings should not be positioned closer than one-sixth of their width from the edge of the wall or one-sixth of their combined length from each other.

There are special constructional details required around the openings at the following points:

- the top of the opening or head
- the sides of the opening or reveals
- the bottom of the opening or cill.

The head

A *lintel* or an *arch* must be incorporated into the wall in order to carry the loads, imposed by the wall above, over the opening to suitable supports or bearings on either side of the opening.

The lintel

The lintel must be capable of carrying the load imposed upon it without excessive deflection and therefore acts as a type of beam spanning across the top of the opening. The length of the bearing of the lintel onto the loadbearing wall on each side is dependent on the size of the opening being spanned. Where the clear span of the opening (the distance between the *reveals*) is 1200mm or less, the length of bearing should be a minimum of 100mm on each side. For openings with a clear span greater than 1200mm, the bearing length needs to be increased to 150mm.

Lintels may be manufactured from precast concrete, pressed steel or mild steel flat or angle section (see Figures 9.6–9.9). Nowadays pressed steel lintels are favoured more than precast concrete lintels because the latter are much heavier and conduct heat easily from the interior to the exterior of the wall and will therefore contravene the requirements of the Building Regulations Part L with regard to heat loss from the building through thermal bridging. However, manufacturers of autoclaved aerated concrete blocks are now producing lintels in the same material. These are much lighter than the conventional precast concrete lintel and have a better thermal resistance value, thus reducing the effect of thermal

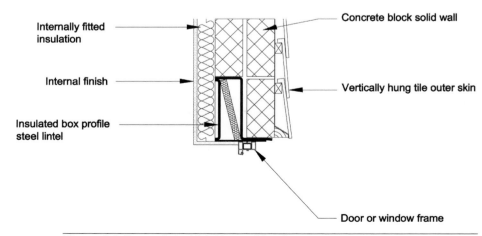

Internally fitted insulation

Internal finish

Insulated box profile steel lintel

Concrete block solid wall

Vertically hung tile outer skin

Door or window frame

Figure 9.6 Pressed steel lintel to a solid wall

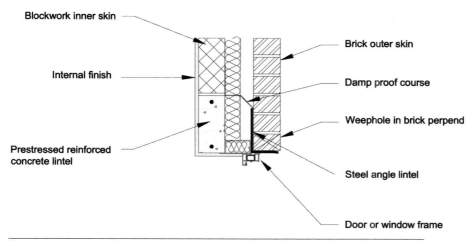

Figure 9.7 Precast concrete lintel and steel angle lintel to a cavity wall

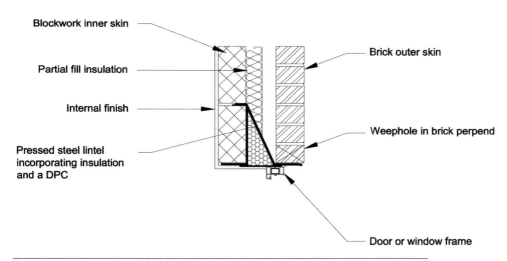

Figure 9.8 Pressed steel lintel to a cavity wall

bridging. Where heat loss through the wall is not an issue (e.g. in the exterior walls of detached garages) then precast concrete lintels may still be used.

In cavity walls a DPC must be incorporated with the lintel to prevent rainwater from crossing the cavity via the lintel. This should be stretched diagonally across the cavity with the inner face being at least 150mm higher than the outer face. Weepholes should be left in the vertical mortar joints or *perpends* between the bricks in the external leaf to drain any water from the lintel out of the cavity. For this reason pressed steel and concrete lintels are usually sloped from their inner to their outer face, giving them the appearance of a boot (for this reason they are sometimes referred to as boot lintels). Because steel is impervious to water, the pressed steel lintels already incorporate a DPC, unlike precast concrete lintels, but it is often advisable to place a shaped cavity tray DPC above them.

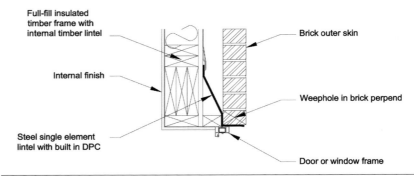

Full-fill insulated timber frame with internal timber lintel

Internal finish

Steel single element lintel with built in DPC

Brick outer skin

Weephole in brick perpend

Door or window frame

Figure 9.9 Steel lintel used in a timber framed construction

Self-assessment question 9.3

Why should the inner section of a boot lintel be taller than the outer section?

In order to prevent heat from escaping across the cavity through the thermally conductive steel lintel, the hollow centre of these lintels is generally filled with a material having a high thermal insulation value, such as expanded plastic.

The arch

In solid walls an arch may be used in preference to a lintel. These are basically of two types:

- rough arch
- gauged arch.

The principle of the arch is that all the bricks forming each side of the arch are placing a compressive force against the central brick in the arch or *keystone*. If the keystone in the arch were to be removed, the arch would collapse. Arches are curved in shape and generally have the profile of a segment to a circle (see Figures 9.10 and 9.11).

However, arches may also be constructed with only a slight cambered profile and appear almost flat in shape (see Figure 9.12).

The rough arch

This arch is curved but the bricks are uncut, leaving wedge-shaped mortar joints between the bricks. To prevent the joints becoming too large, the arch is usually constructed in headers (see Figure 9.10). It does not have the pleasing appearance of the gauged arch, but it is cheaper to construct.

The gauged arch

This arch is constructed from wedge-shaped bricks called *voussoirs*, which produce a uniform thickness of mortar joint between them throughout the arch (see Figure 9.11). The shape of each voussoir is marked out from a template before the brick is cut.

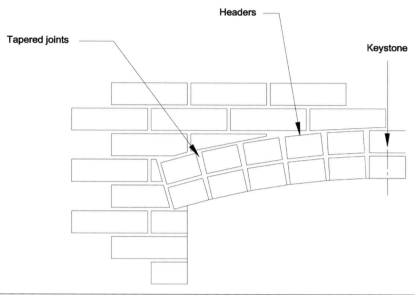

Figure 9.10 Segmental rough arch

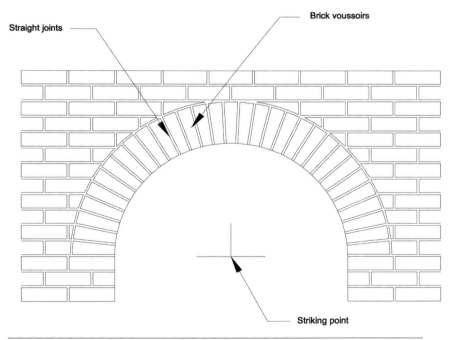

Figure 9.11 Semi-circular gauged arch

Arch construction

When constructing an arch it must be given temporary support until the mortar joints have set and the arch has gained sufficient strength to carry the load over the opening. These temporary supports are called *centres*. They are usually made from timber, and the span, the profile and the load of the arch being constructed govern their design.

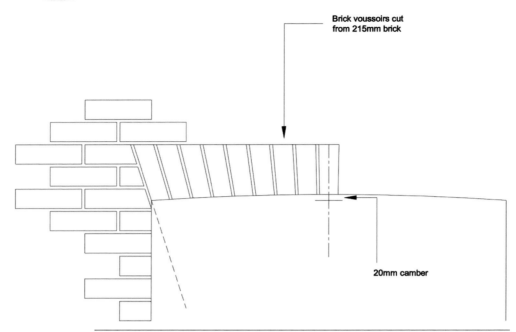

Figure 9.12 Cambered arch

The centre consists of a frame cut to the correct profile of the arch and upon which are secured thin laths of timber to support the brickwork of the arch. The centre is held firmly at the required height by braced props standing on *sole pieces*, which provide a stable base for the props and allow the dead load being carried by the props to be transferred safely to the ground beneath. Fine adjustment of the height of the centre can be made with the aid of *folding wedges* (similar to those used to tighten the struts between the walings in planking and strutting earthwork supports). Alternatively, telescopic metal props may be used (see Figure 9.13).

Arched cavity tray

For cavity wall construction, arched cavity trays are available. They are similar in construction to pressed steel lintels (see Figure 9.14).

The reveals

In solid wall construction the brickwork in the reveals is bonded to provide adequate strength through the wall.

In cavity walls the cavity should be closed at the reveal. This may be achieved by building the blockwork used to construct the inner leaf across the cavity to meet the brickwork outer leaf at this point. However sufficient insulation needs to be provided in order to overcome thermal bridging at this point.

Where blockwork is used to close the cavity at the reveal an insulated vertical DPC must be incorporated between the brickwork and blockwork where the cavity is closed (see Figure 9.15). This is particularly important

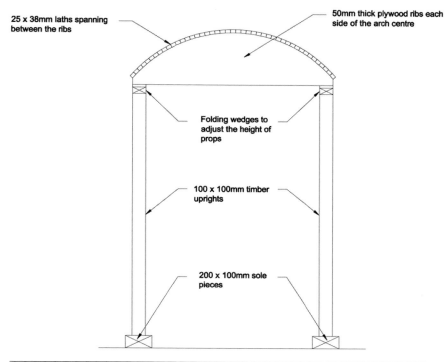

25 x 38mm laths spanning between the ribs

50mm thick plywood ribs each side of the arch centre

Folding wedges to adjust the height of props

100 x 100mm timber uprights

200 x 100mm sole pieces

Figure 9.13 Arch construction

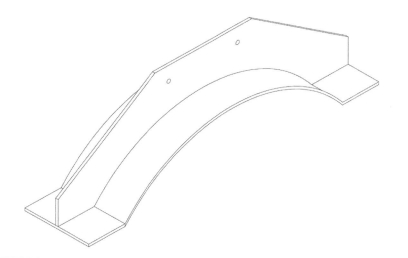

Figure 9.14 Arch cavity tray

where partial fill insulation is used, in order to maintain thermal continuity and decrease the chance of thermal bridging. In circumstances where it is decided not to return the inner leaf across the cavity then cavity closers can be used to close the gap. The benefits of the cavity closer are that it is fully insulated to prevent against thermal bridging and it has fins built in to prevent the passage of water across the cavity. The cavity

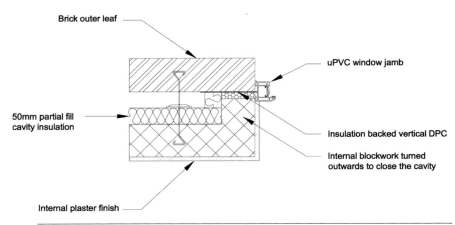

Brick outer leaf

uPVC window jamb

50mm partial fill cavity insulation

Insulation backed vertical DPC

Internal blockwork turned outwards to close the cavity

Internal plaster finish

Figure 9.15 Reveal to a cavity wall

closer can be installed at the side reveals, the head and the cill position to close off the cavity. A number of different styles of cavity closer are now manufactured which can be installed during the construction of the opening or fitted at a later stage once the opening is completed. Cavity closers are manufactured in a number of widths to suit most cavities. Where the width of the *jamb* of the window or door frame may not be wide enough to completely cover the cavity closer, then the reveal may be lined with plasterboard or a thermal backing board (see Figure 9.17) To strengthen the wall around the reveal, wall ties should be positioned within 225mm of the reveal and at 300mm vertical centres for the entire height of the opening (see Figures 9.16 and 9.17).

Self-assessment question 9.4

Why does a DPC need to be incorporated at the reveal where blockwork is used to close the cavity?

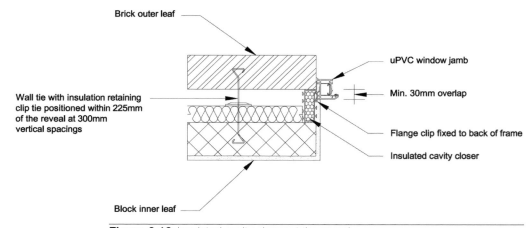

Brick outer leaf

uPVC window jamb

Min. 30mm overlap

Wall tie with insulation retaining clip tie positioned within 225mm of the reveal at 300mm vertical spacings

Flange clip fixed to back of frame

Insulated cavity closer

Block inner leaf

Figure 9.16 Insulated cavity closer at the reveal

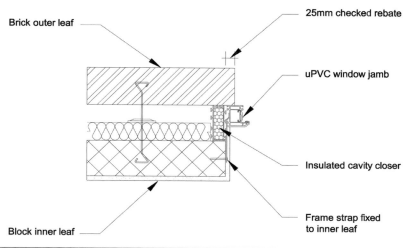

Brick outer leaf

25mm checked rebate

uPVC window jamb

Insulated cavity closer

Frame strap fixed
to inner leaf

Block inner leaf

Figure 9.17 Checked rebate

In areas where exposure to driving rain is excessive (zone 4 as shown in Diagram 12 of the Building Regulations Approved Document C), the window or door frame needs to be set back behind the outer leaf of masonry to form a checked rebate, which should overlap the frame by a minimum of 25mm (see Figure 9.17). Alternatively, in exposure zone 4, an insulated finned cavity closer may be used, removing the need for a checked rebate.

The cill

The function of the *cill* is to shed the rainwater that has run down the face of the door or window above, away from the wall below. The cill may be part of the door or window frame (in the door this component is normally called the *threshold* and will be considered in more detail in the later chapter on doors) or may be a separate component manufactured from precast concrete or stone. A brick cill may also be constructed from bricks having a sloping or cambered top edge to help shed the rainwater from the wall below (see Figure 9.18).

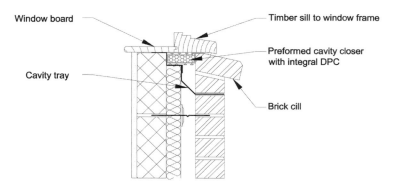

Window board

Timber sill to window frame

Preformed cavity closer
with integral DPC

Cavity tray

Brick cill

Figure 9.18 Brick cill to a cavity wall

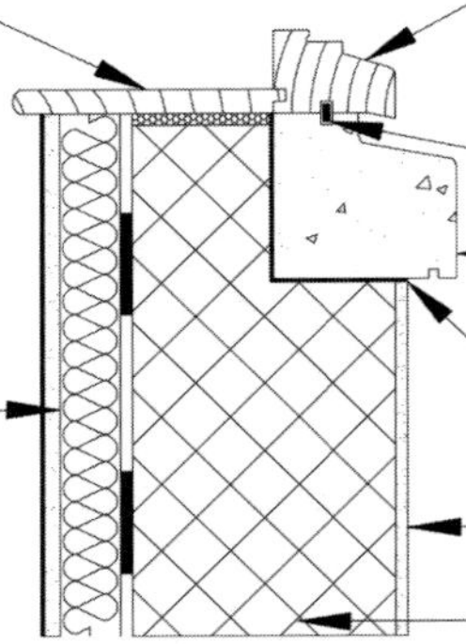

Window board

Timber sub-sill

Galvanised mild steel water bar

Precast concrete cill

Insulated plasterboard
on adhesive dabs

Damp proof course

10mm polymer render

215mm external block wall

Figure 9.19 Precast concrete cill with water bar to a solid wall

Where the cill member bridges the cavity it must have a DPC incorporated beneath it, which is then cloaked up behind its back. To discourage water penetration across the underside of the cill it is *throated* with a small groove on its underside. To prevent water penetration across its flat top edge the cill can incorporate a galvanised mild steel *water bar* to link it to the bottom member of the door or window frame and act as a barrier to water attempting to cross this section of the cill. Again it is necessary to ensure that heat is not lost across this junction and so a cavity closer incorporating thermal insulation to overcome the thermal bridge must be used (see Figure 9.18).

Self-assessment question 9.5

Why does a throat discourage water penetration across the underside of the cill?

To cover the inner part of the wall at the cill in a window opening a *window board* is normally joined to the back of the bottom section of the window frame.

Visit the companion website to test your understanding of Chapter 9 with a multiple choice questionnaire, and to see the following features illustrated with full-colour photos:

- **9.1** Brick cill to external wall
- **9.2** Reconstituted stone cill
- **9.3** Segmental rough arch
- **9.4** Timber frame with concrete block outer skin

CHAPTER 10

Timber

Introduction

Wood is an organic material formed by the growth of cells in living trees. The trees produce this organic matter by the conversion of *sap* to *carbohydrates*. This process is aided by *photosynthesis*. Foodstuffs are consumed in the *cambium* to form living cells containing *protoplasm*. These cells subsequently modify and *lignify* to form the different classes of wood cells. Wood cells are composed mainly of *cellulose* and the cementing agent *lignin*. Most woods are slightly acidic.

Timber is a very useful material for building as it is readily available, relatively cheap, easily converted into the shapes and sizes required, relatively strong and has an attractive appearance. It is also a renewable resource, especially softwoods from northern temperate forests. Timber is used in building for structural members such as beams, joists and rafters, for *carcassing* purposes such as providing the skeleton framing for timber partitions and ductwork, for manufacturing purposes such as the making of doors, windows and stairs and for joinery purposes such as *skirtings*, *architraves* and decorative panelling.

This chapter considers the structure of timber, the methods used to convert the timber into useful sections for construction purposes, the methods of *seasoning* the timber, methods of grading the timber, common defects found in timber, the physical properties of timber, consideration of the causes and remedies of fungal attack and insect attack, methods of improving the durability of timber by preservative treatment and the manufacture and use of timber products such as plywood.

Classification

Timber may be roughly classified into two broad categories:

- Hardwoods
- Softwoods.

Hardwoods differ from softwoods in their cellular structure and thus exhibit different qualities and characteristics. Softwoods have only two types of cell, the pod-like *tracheids,* which form mechanical tissue for

Table 10.1 Typical imported and home produced timbers

Typical hardwoods	Typical softwoods
Imported Afrormosia, Mahogany, Obeche, Sapele, Teak, Utile.	*Imported* Douglas Fir, Parana pine, Pitch Pine, Redwoods, Western Hemlock, Western Red Cedar.
Home produced Ash, Beech, Birch, Elm, Oak, Sycamore.	*Home produced* European Spruce, Larch, Scots Pine, Yew.

strength and cells for conducting water and foodstuffs through the tree, and the brick-shaped *parenchyma* cells that provide food storage. Hardwoods, on the other hand, possess needle-like fibres as mechanical tissue and tube-like vessels providing cells conducting foodstuffs in addition to the parenchyma and sometimes the tracheids.

Hardwoods are generally very strong and durable and, because of their fine grain pattern, are often used for decorative purposes. Most hardwoods come from broad leafed, *deciduous* trees, although tropical hardwoods come from evergreen varieties. Hardwoods are generally more expensive than softwoods, although cost varies with species and the dimensions of timber required.

Softwoods come from *coniferous* trees that are mainly evergreens and grow in northern temperate forests. They often contain resins that may interfere with paint coatings. In general they are quick growing and so are generally cheaper than hardwoods and comprise about 75 per cent of the timber used in the U.K. construction industry, particularly for structural and carcassing uses. However they have a lower density than hardwoods and generally have lower strength. They are also often less durable than hardwoods and may therefore require preservative treatment.

Structure

A tree is made up of the following elements:

- *Growth rings* – As the tree grows it produces small tubular cells in the form of rings. Each ring has two parts: *Springwood* – a wide range of fibrous cells and *Summerwood* – a narrow band of dense dark cells. Springwood has a typical density of 350 kg/m^3, approximately half the density of summerwood.
- *Sapwood* – This is formed by new growth rings and is mainly used for conducting and storing foodstuffs.
- *Heartwood* – As the older sapwood is replaced by new wood, it changes its function to that of providing structural stability to the tree and becomes heartwood. The proportion of sapwood therefore decreases as the tree ages. Sapwood is generally lighter in colour than

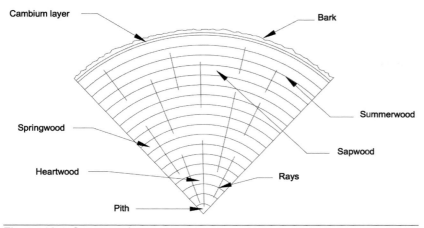

Figure 10.1 Structure of a tree

heartwood and because it contains water, starch and sugars it is more likely to suffer from fungal or insect attack. In consequence, heartwood is expensive.

- *Grain* – This is the general arrangement of fibres in the wood. It is often the characteristic that affects the appearance of a particular piece of timber. Thus decorative timbers are often chosen for their grain structure.
- *Rays* – These are the radially arranged food cells called *parenchyma*. They conduct food transversely through the wood.
- *Pith* – This is the spongy central cellular core to the tree.
- *Bark* – The outer sheath of cork that protects the tree.
- *Cambium* – A layer beneath the bark where cells are manufactured to produce growth rings (see Figure 10.1).

Conversion

This involves the cutting up of a log into sawn timber or *veneers*. In the conversion process various types of saw are used, but principally circular saws, band saws and frame saws.

The log may be cut in various ways but the two most typical are:

- Flat sawn – The log is cut longitudinally into flat boards (see Figure 10.2).
- Quarter sawn – The log is sawn into quarters and each quadrant is subsequently sawn to produce flat boards (see Figure 10.3).

The method used for conversion has a bearing on the resultant grain pattern available and the wear resistance of the timber. Flat sawn timber from softwood has a more decorative appearance than quarter sawn, but this does not generally apply to hardwoods, since their grain pattern depends largely upon irregularities in the grain as well as the arrangement of rays. Quarter sawn boards shrink less than flat sawn boards and

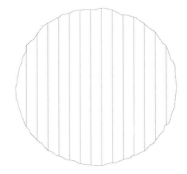

Figure 10.2 Flat sawn timber

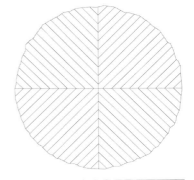

Figure 10.3 Quarter sawn timber

generally give more even wear if used in flooring. However, the waste is greater in quarter sawing than it is in flat sawing.

Seasoning

Timber from a recently felled tree contains moisture in the cells, particularly the sapwood. The amount of moisture contained in the felled tree is dependent on its species and age and can vary from 22 per cent to 60 per cent. Generally softwoods have higher moisture contents than hardwoods. If the timber is installed in the building with this high cell moisture content still present it will, in time, undergo a natural drying process from the temperature and humidity conditions prevailing within the building. As the cell moisture evaporates the timber will begin to shrink, leaving unsightly gaps at joints or even causing *warping* of the timber. Also, timbers containing high amounts of moisture in their sapwood are more prone to fungal or insect attack.

Self-assessment question 10.1

Why do softwoods generally have higher moisture contents than hardwoods when they are freshly felled?

Seasoning is the process of reducing the moisture content within the timber, by controlled methods, to a level appropriate to its final use, before it is fixed within the building. Timber is *hygroscopic*, that is it has the ability to take in or give off moisture vapour until it reaches an equilibrium value with the humidity conditions of the place that it is located. Timber required for internal use should be seasoned such that its eventual moisture content approximates with the average humidity of the room into which it is to be fixed. The moisture content of timber is generally calculated as a percentage of the dry weight of the wood and should be reduced according to the purpose to which the timber is to be put. The moisture content of timber in a well-heated building is likely to be 8 per cent in summer and 15 per cent in autumn, whilst externally the moisture content of timber may typically change from 15 per cent in summer to 22 per cent in winter.

Seasoning may be accomplished by two main methods:

- air seasoning
- kiln seasoning.

Air seasoning

The timber is stacked outdoors, protected by temporary sloping roofs, with gaps between each board and the air is allowed to flow around the timber. The process is slow, especially during winter months, and it is not usually possible to reduce the moisture content much below 17 per cent by this process.

Kiln seasoning

The timber is stacked in a kiln with gaps between each board, and air, heated by passage over steam pipes to the desired temperature, is circulated through the piles. The air must contain a certain amount of humidity, regulated by the admission of steam, to prevent the timber from drying too quickly and splitting. The humidity level of the air is gradually reduced as the timber dries.

This process is much speedier than air drying, is closely controlled and allows the reduction of moisture content to be taken to levels as low as 12 per cent.

Self-assessment question 10.2

What is the main disadvantage with kiln seasoning?

A combination of air and kiln seasoning can be used whereby the moisture content of the timber can be reduced to approximately 20 per cent by air seasoning before further treatment is carried out by kiln seasoning to reduce the moisture content of the timber to the required level. By this method the amount of time the timber must remain in the kiln is reduced and kiln throughput is considerably increased, but the process takes longer than with kiln seasoning alone.

Timber defects

There are a number of defects that can affect the suitability of timber for use in construction:

- *Knots* – created when a part of a branch becomes enclosed in a growing tree. Knots cause problems because of their hardness, causing uneven wear in floorboards and difficulty in working the timber. There are several types of knot. A sound knot is one free from decay and at least as hard as the surrounding wood. A dead knot has its fibres interwoven with those of the surrounding wood. A loose knot is a dead knot that is not held firmly in place. *BS EN 942:2007 Timber in joinery. General requirements* does not allow exposed, decayed or dead knots in joinery timbers, but will allow sound, tight knots provided they are not too large. They are often considered to be unsightly in hardwood, but can enhance the appearance of softwood. However, they often contain resin, which must be sealed before the wood is painted, otherwise it may exude and damage paint coatings. Knots, because they interrupt the bonding of the natural fibres of the timber, often reduce the timber's strength in tension but not in its resistance to shear and splitting.
- *Rind galls* – surface wounds that have eventually become enclosed in growth and thus affect the continuity of the natural fibres within the timber.
- *Burrs* – swellings of highly contorted grain, resulting from many undeveloped buds that have formed over a wound. They do not affect

the strength of a timber and are often used to decorative effects in oak and walnut veneers.

- *Shakes* – a separation of fibres along the grain of the timber, which are caused by stresses within the living tree. A cross shake occurs in cross grained timber and follows the grain. A heart shake occurs radially originating at the heart of the tree. A ring shake follows a growth ring.
- *Brittle heartwood* – if the heartwood is too brittle it may suffer from brittle fracture when subjected to a load.
- *Wide growth rings* – this indicates too rapid growth, resulting in thin walled fibres, with a consequent loss of density and strength.
- *Spiral grain* – this may lead to distortion of the timber during seasoning.
- *Splits* – a separation of fibres along the grain of the timber forming a crack or fissure that extends through the piece of timber from one surface to another. This reduces the timber's usability, since they affect its strength and appearance. Splits are often caused by too rapid drying in the kiln.
- *Checks* – similar to splits but the crack does not extend through the timber from one surface to another. Again too rapid drying in the kiln often causes these.

Strength grading

Timber can be graded to determine whether it is suitable for structural purposes. This strength grading was initially done by visual inspection and considered the defects in the timber that could affect its structural strength. Visual inspection is covered by *BS 4978:2007 + Amendment 1:2011 Visual strength grading of softwood. Specification*, and relies on defects in the timber which could affect its strength being apparent on its surface. The structure of the timber and its density may not be sufficiently taken into consideration. It is also a much slower process than machine grading.

Strength grading by machine is covered by *BS EN 14081-1:2005 + Amendment 1:2011 Timber structures. Strength graded structural timber with rectangular cross section. General requirements*. Lengths of timber are passed through the grading machine, which imposes a load at various places along the length of the plank, usually about its minor axis, which is its weakest point, to stress the timber and measures the reaction to that load. Using a correlation between stiffness and strength the timber is graded into strength classes as defined by *BS EN 338:2009 Structural timber. Strength classes*. A strength class is a group of divisions of timber strength into which timber species and strength grade combinations, having similar strength and mechanical properties, are allocated. Timber is classified according to its type and this is prefixed with a C for softwood and D for hardwood and the characteristic bending stress of the timber. The most commonly used strength classes for softwood in building are C16 and C24. Timbers destined to be used in the manufacture of *trussed rafters* are prefixed with the letters TR.

Physical properties of timber

Because the term timber covers such a wide diversity of different species of tree, it is extremely difficult to generalise about the physical properties of timber. However, it is possible to summarise about how certain physical characteristics found in timber have an effect on the physical properties of the timber. The following physical properties will be considered:

- strength
- moisture and thermal movement
- thermal resistance
- sound insulation
- fire resistance
- durability
- appearance
- sustainability.

Strength

Timber has a high strength:weight ratio, that is, it is quite strong both in tension and compression relative to its weight. The strength of timber increases virtually in direct proportion to its density. The tensile strength of timber is up to thirty times stronger along the grain than it is across it and the compressive strength is up to three times stronger along the grain than across it. The strength of timber decreases with its moisture content. Increases in the temperature of timber can also detrimentally affect its strength. Timber is also quite an *elastic* material, that is, it will deform under load but can retain its original shape once the load has been removed.

Strength is also affected by the amount of defects in the timber such as knots, shakes, checks, splits, brittle heartwood and wide growth rings.

Moisture and thermal movement

Because timber is hygroscopic it can easily take up or lose moisture from and to the surrounding atmosphere, causing the timber section to correspondingly swell or shrink. Timber may therefore swell by up to 6 per cent in the tangential direction, by up to 4 per cent in the radial direction but only 0.1 per cent in the longitudinal direction. This shrinkage and swelling movement can have a profound effect on timber, particularly if it has a relatively high moisture content before it is incorporated into a building and subsequently dries fairly rapidly in the heated conditions of the building. Gaps can open up in the joints between members and, if the timber member is restrained, any moisture movement can lead to unacceptable stresses being imposed on the member which may cause it to twist or warp and, in the worst cases, these stresses may cause the timber to crack or split.

Thermal movement is relatively negligible, being slightly more conspicuous across the grain than along it. It should, however, be borne in mind that a rise in temperature of the timber can cause the member to lose moisture, dependent on the humidity level of the surrounding air, and this loss of moisture could cause the timber to shrink.

Thermal resistance

Timber is a good thermal insulator. The thermal conductivity of a typical timber having a density of 481kg/m^3 with a moisture content of 20 per cent is 0.14W/mK. Generally the thermal conductivity of timber increases with density and moisture content.

Sound insulation

Like most materials the sound insulation properties of timber improve with density.

Fire resistance

Although timber is combustible due to the emission of flammable gases when the temperature of the timber reaches about 300^{0}C, it is possible to provide an acceptable level of fire resistance for a timber member being used for structural purposes in a building by the process of *sacrification*. This technique is based on the knowledge that when timber begins to burn it does so from the outside. As the exterior of the timber burns it *chars* (i.e. it produces charcoal). Although charcoal burns it does so at a higher temperature than that of timber (about 500^{0}C) and this layer of charcoal can insulate the timber beneath it as well as prevent oxygen from reaching the internal timber. The consequence of this is that the rate of burning is slowed down. The charring rate of timber is dependent on its density and moisture content but can be estimated as approximately 30mm per hour for timber having a density greater than 650kg/m^3 and about 25 per cent greater for timbers having a density less than 650kg/m^3. Therefore it is possible to determine an acceptable fire resistance for a structural timber member and add on an extra thickness of sacrificial timber to the basic size of the timber member to account for the charring rate.

Self-assessment question 10.3

A structural timber member is required to be 300mm deep x 150mm wide. The timber has a density of 500kg/m^3. It is required to have a fire resistance of 1 hour. Determine the actual size of the member.

Structural timbers used in domestic house construction will normally only be required to achieve a fire resistance of 30 minutes.

Durability

The durability of timber varies with the species and the proportion of heartwood to sapwood in the timber itself. Timber has good resistance to sunlight and frost and can be submerged in water for long periods of time without any adverse effect. The combination, however, of water and air leads to an environment in which *wet rots* may flourish. The greatest risks to the durability of timber are fungal attack and insect attack, which will be discussed in more detail in the requisite sections below.

Timbers may be classified according to their natural durability derived from the amount of natural acids and other chemicals that are present in the heartwood and are toxic to fungi. The *Building Research Establishment Digest DG429 Timbers: their natural durability and resistance to preservative treatment* grades the natural durability of timber by species into five grades (see Table 10.2).

Timbers having low natural durability may have their durability enhanced by the addition of preservatives. Again the ability of timbers to absorb preservative treatments varies with the species. BRE Digest DG429 classifies these into four groups:

- Group 1 – Permeable – amenable to treatment and can be penetrated completely under pressure without difficulty.
- Group 2 – Moderately resistant – fairly amenable to treatment.
- Group 3 – Resistant – difficult to treat.
- Group 4 – Extremely resistant – virtually impervious to treatment.

Fortunately there is a significantly high correlation between perishability and permeability to preservative treatment, although there are some exceptions.

Preservative treatment of timber is considered in a separate section below.

Table 10.2 Durability grades of timbers

Durability grade	Average life of 50 x 50mm stakes in ground contact	Typical species
Perishable	Less than 5 years	Ash, Beech, Birch, Sycamore
Non-durable	5–10 years	Elm, Douglas Fir, Obeche, Spruce
Moderately durable	10–15 years	Cedar, Larch, Mahogany, Sapele, Walnut
Durable	15–25 years	Chestnut, Oak, Utile, Yew
Very durable	Over 25 years	Afrormosia, Ebony, Iroko, Teak

Adapted from *BRE Digest DG429 Timbers: their natural durability and resistance to preservative treatment*, 1998. Reproduced by permission of BRE.

Appearance

Appearance is only really significant in timber used for joinery purposes, as structural timbers and timbers used for carcassing are not generally seen. The appearance of timber is affected by grain pattern and colour along with natural defects such as knots, shakes, rind galls and burrs. In some cases these defects contribute to enhancing the appearance of the timber: for instance, knots in softwood and burrs in some hardwoods. In other cases these defects can detrimentally affect the appearance of the timber and its use for joinery purposes.

Sustainability

The majority of timber used in the UK construction industry is imported, even though many species of softwoods that could be used for construction are home grown. The use of home-grown timber is environmentally and economically preferable to the use of imported timber. The cost of transportation of imported timber from the forest to the building site uses up to 15 times more energy than that used for home-grown timber. However, although the UK also produces hardwoods, many hardwoods specified for joinery applications are derived from tropical forests. There have been concerns over the mismanagement of these forests and so a certification scheme managed by the Forest Stewardship Council (FSC) has now been instigated. Specifiers are now being urged to use FSC timber wherever possible, to ensure it has come from well-managed forests and is a truly sustainable source.

Timber has the highest sustainability credentials of all construction materials. It is a renewable resource and, rather than producing CO_2 in its production, actually takes up CO_2 from the air. It is also recyclable and biodegradable. Reclaimed timber can also be used as floorboards and in studwork, whilst chippings and sawdust produced in the conversion of logs into timber members can also be used in the production of particleboard.

Fungal attack

Fungi are a group of simple plants that do not have leaves or flowers and so cannot manufacture their own food by photosynthesis. They therefore rely on being able to consume stored foods in timber by the action of enzymes. As such they are parasitic organisms. They mainly consume the cellulose in timber, but some fungi will also consume the lignin. In order to digest this material the fungi require the timber to be damp (a minimum moisture content of 20 per cent), the air to be still and a minimum of sunlight. Some fungi will thrive on living trees, others thrive on unseasoned wood. It is intended here to only consider those fungi that attack timber in buildings.

Fungal attack in buildings can be classified into two main types:

- dry rot
- wet rot.

Dry rot

Serpula Lacrymans, formerly known as *Merulius Lacrymans,* is a fungus that attacks the cellulose in timber, breaking the internal part of the timber member into small cubes leaving the wood feeling dry and crumbly.

Although this fungus is often referred to as dry rot, this name can sometimes be misleading as a relatively high moisture content of 20 per cent is needed for this type of rot to occur and thrive. This level of moisture content can often be attributed to the ingress of water through the envelope of the building. The timber then takes up this extra moisture, thus raising its own moisture content.

The life cycle of this fungus is similar to that of other fungi. It begins with the fruiting body releasing millions of spores into the air. These are carried by air currents until they alight on a piece of timber that has the right conditions for growth. The timber must have a moisture content greater than 20 per cent and it should ideally be in a dark position with little ventilation. Because of this, timbers in suspended timber ground floors are particularly susceptible to attack by this fungus. Once the spores have landed on the timber they germinate, fine white strands called *hyphae* spread out across the timber and feed on the cellulose. They then combine to form a cotton wool like structure called *mycelium*. Further strands called *rhizomorphs* develop from the mycelium conveying moisture to new wood and making it suitable for attack. These may spread through plaster and even brick walls to find other timber to attack. In fact it is not uncommon for moisture to be drawn out of damp walls to sustain the fungus. Finally the fruiting body called a *sporophore* is produced and when conditions are favourable it releases spores into the atmosphere and the cycle commences again.

Self-assessment question 10.4

Why are timbers containing higher proportions of sapwood to heartwood particularly vulnerable to fungal attack?

The symptoms of an attack of dry rot are a characteristic 'musty' smell, the appearance of spreading mycelium, rhizomorphs or the fruiting body (which is whitish-grey when it first appears and then becomes leathery), the appearance of deep, cube-like cracks across the grain of the timber and finally a total loss of strength and density in the timber.

Remedial treatment has to begin with a survey of the building to determine the extent of the attack. This will require the removal of floorboards and may require the removal of plaster to walls. All infected timber must then be removed and burnt. The area where the fungus was established then needs to be treated with a liberal application of fungicide. Affected walls should be treated with a blowlamp to kill any fungus that may still be present. The cause of the infection should then be tackled; this may be a leaking gutter or down pipe, a broken roof tile or defective flashing or a lack of adequate ventilation beneath a suspended timber ground floor, perhaps caused by blocked air bricks. Finally the affected timbers that have been destroyed must be replaced with new timbers that have been adequately treated with preservatives.

Wet rot

This is a general name given to a group of fungi that are of a similar type. *Coniophora Puteana,* formerly known as *Coniophora Cerebella* and also known as cellar fungus, is the most common. These fungi need a moisture content in the timber of at least 25 per cent. This is generally too wet for dry rot to survive.

Wet rots can normally be identified by fine dark brown or blackish strands and a green leathery fruiting body can sometimes be seen. The timber develops deep cracks along the grain and the infestation is normally localised to the area of saturated timber.

Phellinus Contiguus, formerly known as *Porio Contigua*, is a wet rot that attacks and consumes both the cellulose and the lignin in timber, particularly windows and door frames. The affected timber breaks into soft strands.

Because wet rots do not normally spread to neighbouring dry timber and brickwork, as is the case with dry rot, the remedial treatment is to cut out the affected timber, treat the area with fungicide, remedy the cause of the dampness and replace the affected timber with suitably preserved timber.

Modern techniques being developed by biologists in the war against fungal infestation are the propagation of enzymes that cause fungi to consume themselves, rather than the timber.

Insect attack

Most of the insects responsible for damage to timber in buildings are beetles. They have a characteristic life cycle. The adult insects lay their eggs in cracks or crevices in the timber. The larvae develop and bore into the timber, feeding on the cellulose and leaving behind a fine powder called *frass*. It is this stage that causes the greatest damage to affected timbers. After a period of time that alters with the type of insect (from one to five years), the larvae make their way to just below the surface of the timber and turn to pupae; after a few weeks in this stage the grubs then emerge as fully developed adult insects. They make their way to the surface of the timber and emerge through flight holes to mate and repeat the cycle.

Insect attack may be identified by the appearance of the adult beetles, the appearance of the larvae, the appearance and texture of the frass and the size, shape and cross section of the bore tunnels. Although there are a wide number of insects that can attack timber in buildings the most commonly found are:

- Death Watch Beetle
- Common Furniture Beetle
- House Longhorn Beetle.

Death Watch Beetle

These are 6 to 8mm in length and dark chocolate brown in colour. The white, lemon-shaped eggs are laid in clutches of approximately 70 in cracks and crevices in hardwoods, especially old timber, often after fungal decay has taken place. The larvae are curved in shape and create a bore dust of bun-shaped pellets that are visible to the eye. The larval stage normally lasts for 3–10 years and the pupal stage normally lasts for 3–6 weeks. They then produce a characteristic rapping of their heads against the timber as a mating call and then emerge through 3mm diameter flight holes in the spring.

Common Furniture Beetle

These are 2–3mm in length and reddish to blackish brown in colour. The beetle, often known as woodworm, attacks the sapwood in timber. The larvae are crescent-shaped and create a bore dust with cigar-shaped pellets that is gritty when rubbed between the fingers. The larval stage normally lasts for 1–3 years and the beetle emerges from 1.5mm diameter flight holes in July or August.

House Longhorn Beetle

These are 8–25mm in length and brown or black in colour. The beetles only attack seasoned softwoods and are mainly confined to the south-east of England. Building Regulations Approved Document A, Table 1 specifies the geographical areas of the country where softwood timber for roof construction or fixed within the roof space should be adequately treated to prevent infestation. The larvae are straight bodied and white in colour and create a very coarse bore dust, having tiny chippings contained within it. The tunnels are often 12mm wide and shallow. Flight holes are oval-shaped and 6–12mm in diameter.

Rectification of insect attack needs to commence with a survey to determine the extent of the attack. Powerful vacuum cleaners are then required to remove all dust and frass from the timber prior to liberally treating the affected timbers with insecticide. Badly infected timber should be removed and replaced with new timbers that have been treated with preservatives.

The use of harmful chemicals to treat insect attack is being replaced in some parts of the world now by the use of irradiation of affected timbers or exposing the timber to ultra-violet light. Biologists are finding anti-feedants that put larvae off their food and are attempting to synthesise the sex-allure scent of the female Common Furniture Beetle to lure the male into a sticky trap before it can mate.

Preservative treatment

As discussed in the section on durability of timber, timbers of differing species vary considerably in their natural durability. Although it is advantageous to use timbers that have high natural durability, it is not always economic to do so. Timbers with low natural durability may be used in situations where decay may occur, provided they have received initial treatment with a preservative.

Decay may be prevalent in the following situations:

- where the timber is in contact with the ground
- where the timber is used at or below DPC level
- where the timber is wholly enclosed in masonry or concrete
- where adequate ventilation cannot be provided
- where the equilibrium moisture content of the timber is likely to exceed 20 per cent
- where low durability timber is used in high risk situations where temperature, ventilation and humidity levels are unacceptable
- where fungal and insect attack are prevalent.

The end use of timber is classified by *BS EN 460:1994 Durability of wood and wood-based products. Natural durability of solid wood. Guide to the durability requirements for wood to be used in hazard classes*. There are five hazard classes, which are based on the risk of fungal or insect attack. These hazard classes determine the type and degree of treatment (see Table 10.3).

Timber preservatives should have the following qualities:

- easy to apply
- able to penetrate deeply into the timber
- highly toxic to fungi and wood boring insects at relatively low concentrations

Table 10.3 Hazard classes for timber

Hazard class	Hazard
1	Internal with no risk of wetting nor condensation
2	Internal with risk of wetting or condensation
3a	External above DPC – coated
3b	External above DPC – uncoated
4	External, soil or fresh water contact

Adapted from *BS EN 460:1994 Durability of wood and wood-based products. Natural durability of solid wood. Guide to the durability requirements for wood to be used in hazard classes*. Reproduced by permission of BSI.

- able to be retained in the timber, even when saturated
- no smell or colour
- harmless to operatives, residents, paint coatings and furnishings
- non-corrosive to metals
- non-inflammable
- available at a reasonable cost.

No one preservative can claim to have all the qualities listed above but the best have most of them. However, careful thought needs to be given to whether timber preservatives really need to be applied, as they can contain hazardous and carcinogenic substances that can be harmful not only to the environment but also workers who are involved in manufacturing and applying them. Treated timber is classified as hazardous waste and its disposal can threaten the environment. *BS 8417:2011 Preservation of wood. Code of practice* gives recommendations and guidance for the preservative treatment of wood. It gives recommendations for the need for treatment and specifies the type of treatment.

Timber preservatives are classified into four distinct types:

- tar oil preservatives
- waterborne preservatives
- light organic solvent preservatives
- micro-emulsion preservatives.

All four provide protection against fungal decay and attack by wood boring insects.

Tar oil preservatives

The best known of these is creosote, covered by *BS 144:1997 Specification for coal tar creosote for wood preservation*. However tests on this material have shown it to be carcinogenic and so, since 2003, it can only be used under controlled conditions and not applied on site.

Tar oil preservatives are leach resistant (they do not easily drain from the wood when saturated with rain water) and so are particularly suitable for treating timbers used in exterior work, immersed in water or buried. They can be applied to timber with a relatively high moisture content. They are not usually corrosive to metals. However, they are difficult to paint over and the preservative tends to creep when placed in contact with plaster or other absorbent materials. They have a strong odour, which may be picked up by food placed nearby. Freshly treated timber is more flammable than untreated timber, due to the volatile oils used as a solvent, but the flammability of the treated timber is no greater than that of untreated timber once these volatile oils have evaporated.

Waterborne preservatives

These are cheaper and easier to transport than other preservatives as they are in powder form and can be mixed with water on site. However, water is not absorbed quickly into timber and so waterborne preservatives are normally applied by processes that overcome the timber's natural resistance to penetration. This is typically vacuum impregnation under high pressure.

Copper has been the main ingredient for many of these preservatives, as it is effective against wood-destroying fungi. To widen the effectiveness against a range of fungi and insects, it is generally mixed with other compounds such as arsenic, boron or phosphates. Chromium compounds are added to the mixture to provide a fixative to improve the resistance of the preservative to leaching. For many years the most widely used waterborne preservative was copper chromium arsenic (CCA). However, since 2004 its use has been restricted and it is no longer permitted for use in residential buildings. It has been replaced by chromated copper boron (CCB). However, the safety and environmental impacts of chromium are also causing concern and there may well be restrictions on this type of preservative in the future.

Alternative treatments using copper with a range of organic active ingredients such as azoles and quaternary ammonium compounds are now available. The most environmentally friendly of these formulations however uses inorganic borates. Boron is easily diffused into timber and boron plugs can be inserted into pre-drilled holes in affected timber to arrest the infestation.

Waterborne preservatives are odourless, non-inflammable, non-creeping and easily penetrate into most timbers. The treated timber generally has a light-green colouration, which is not a problem on structural timbers that are not exposed to view when installed into the building and enables users to determine that the timber has been treated. They may be painted over when the treated wood has dried, if desired. However, although they are non-leaching initially, there is a tendency for the preservative to leach in time when used on timber in contact with the ground or water, particularly if chromium salts have been omitted. They can also be corrosive to metals, and re-drying of the timber after treatment is necessary.

Light organic solvent preservatives

These are most often applied by double vacuum pressure. The preservative chemicals, typically Permethrin as an insecticide and azoles as fungicides, are dissolved into a light organic solvent, such as white spirit, which takes the preservative into the timber and is then drawn out in the last stages of the double vacuum process. They are covered by *BS 5707:1997 Specification for preparations of wood preservatives in organic solvents*. These are non-corrosive to metals, non-creeping, leach resistant and penetrate well into most timbers. Unlike waterborne preservatives they do not swell the timber on application, and re-drying of timber after treatment is not required. The treated timber is colourless and may be painted when dry. The preservatives are non-injurious to plant life after the solvent has evaporated. However, there is a need for timber to receive a protective coating when these preservatives are used in an exterior environment and they are now generally confined to joinery work in hazard class 1, which could be affected by the raising and swelling of the grain when waterborne preservatives are used. Similar to tar oil preservatives, a volatile solvent is used for these preservatives, so the freshly treated timber is more flammable than untreated timber until the volatile solvent has evaporated. They also have a strong odour that may be picked up by food placed nearby and are the most expensive both to manufacture and to apply.

Micro emulsions

These provide an alternative to waterborne and light organic solvent preservatives, as they offer many of the advantages of each technique and overcome some of the consequent disadvantages. They use the same insecticides and fungicides as light organic solvent preservatives, but high concentrations are dissolved in strong solvents. They are then mixed with emulsifiers, so that the ingredients can be mixed with water. The uptake of the preservative is controlled during the treatment process, so swelling of the timber is much lower than that experienced with normal waterborne preservatives.

The uptake of preservatives by timber is related to the anatomical structure of the wood and the amount of water in the cells. For this reason timber should be seasoned before treatment with any preservative, but particularly where waterborne preservatives are to be used, and all necessary cutting, shaping and boring should be undertaken prior to treatment. However, if complete working of the timber prior to treatment is impossible a liberal application of preservative should be applied to all cut surfaces prior to fixing. As stated before, this last measure cannot be used with creosote, but other substitutes are now available that are less hazardous to health and may be applied in situ.

Application of preservatives

This may be undertaken by four main methods:

- brushing or spraying
- dipping, deluging or steeping
- hot and cold open tank
- pressure impregnation.

The method employed will depend upon the following factors:

- the location in which the timber is to be used
- the type of timber
- the permeability of the timber
- the type of preservative being used
- the cost of the application process.

With brushing or spraying the penetration into the timber is limited. The surfaces to be treated must be effectively coated and the process repeated every 2–3 years.

Deluging, dipping or steeping ranges from a quick dip of the sections to be treated into the preservative (dipping), laying the timber sections in a tank and pouring the preservative over them (deluging), to laying the timber sections in a tank of preservative for several weeks (steeping). As can be expected, of these three methods steeping gives the greatest depth of penetration and dipping the least. These techniques are not used widely nowadays.

The hot and cold open tank method involves immersing the timber into a tank of preservative, which is then heated to 80–90⁰C and then allowed to cool. This forces the preservative into the cells of the timber and gives a better depth of penetration than the cold applied methods previously considered. However it is more expensive as it uses energy to heat the preservative. It also requires the use of equipment that is not available on site. Again this method is not widely used nowadays.

Pressure impregnation involves the timber being sealed in a pressure vessel; air is then removed from the vessel under vacuum. The preservative is then forced in under high pressure. Where light organic solvent preservatives are being used the vacuum process is undertaken under lighter pressure and a second vacuum stage removes any excess preservative. Both techniques may be used either by the full cell method, in which the cells of the timber are filled with preservative or the empty cell method, in which only the walls of the cells remain coated with preservative. The full cell method tends to be more effective with waterborne preservatives. Pressure impregnation is the most expensive application process but gives the best depth of penetration of all. This is particularly useful in treating timbers with low permeability.

Timber products

The sizes of timber members that can be used in a building are normally restricted to the size of the tree from which the timber has been derived. This is rather restrictive for sheet or board products; therefore various methods have been developed in order to manufacture sheets and boards from timber or derivatives of timber. This has been made easier in recent years with the development of strong resin adhesives that are able to bond composite materials together to form sheets or boards up to 1.2 x 2.4m in size that can be used in a number of building applications.

The most common of these sheets or boards are:

- Plywood
- Blockboard and laminboard
- Particleboard
- Fibreboard.

Plywood

Thin strips of timber, called veneers, are stripped off the circumference of logs by a cutter similar to a giant pencil sharpener. These veneers or *plies* are then bonded together to form a board to *BS EN 636:2003 Plywood. Specifications*. In order that movement stresses within the board are balanced the alternate plies are crossed at 90⁰ to each other around a central core ply. This produces an odd number of plies (3 ply, 5 ply etc.) but the resulting board is very stable and extremely strong for its thickness.

Self-assessment question 10.5

How has the plywood board been able to be made strong yet thin?

The strength of plywood and its end use is determined by the bond performance of the adhesive. There are three classifications of bond performance covered by *BS EN 314-2:1993 Plywood. Bonding quality. Requirements*:

- Class 1 for dry interior uses
- Class 2 for high humidity environments, such as covered exterior uses
- Class 3 for exterior uses out of ground contact.

Urea-formaldehyde resin is usually only suitable for Class 1 applications. Melamine urea-formaldehyde resin is usually suitable for Class 2 applications and Phenol-formaldehyde resin is needed for Class 3 applications.

BS EN 636:2003 also classifies plywood for end use using the Class 1, Class 2 and Class 3 categories outlined above. Although the adhesive used is important in these categories, the type of veneer is also important for exterior applications and must be of durable quality. Plywood can be engineered to have specific structural strength characteristics. Structural plywood is denoted by the suffix S after the class category.

Self-assessment question 10.6

Why should the bonding quality of the adhesive affect the end use classification of plywood?

Plywood may also be classified by its surface appearance, particularly when used in joinery applications. *BS EN 635-1:1995 Plywood. Classification by surface appearance. General* sets out the classification system and the grades, which are then considered in 635-2 for hardwood plies and 635-3 for softwood plies. There are five different grades of veneer, with the best, Grade E, having no defects and the worst, Grade IV, having virtually no limits on the type and size of defects.

Plywood is an expensive sheet material and therefore tends only to be used where its enhanced strength, low moisture movement and good durability properties are particularly important. It is mainly used for floor and roof decking, but is also particularly useful for sheathing in timber frame construction. Its high strength to weight ratio also makes it a suitable material for formwork. Plywood with decorative veneers can also be used for panelling as an internal finish and in joinery applications, such as the manufacture of doors.

Blockboard and laminboard

Blockboard comprises a group of timber 'blocks' generally 25mm in width and 12–43mm in thickness glued together to form a solid core onto which is then bonded a ply veneer to each side. The resulting board has the strength equivalent to the timber used for the blocks and is useful for the construction of shelves and cupboards.

Laminboard is a similar product to blockboard but the strips of timber used for the central core are much thinner, generally 7mm. In consequence, the board produced has less surface distortion and therefore produces a higher quality surface finish.

Both blockboard and laminboard are used in solid core flush door construction and other joinery applications. Their manufacture was covered by *BS 3444:1972 Specification for blockboard and laminboard*, which has subsequently been withdrawn and not replaced.

Particleboard

These comprise particles or 'chips' of timber bonded together with synthetic resin adhesives, similar to those used in plywood manufacture. The mixture is then pressed under heated conditions to set the adhesive and to form a board.

There are a wide range of different qualities of particleboard available. These are covered by *BS EN 312:2003 Particleboards. Specifications*, which has also subsequently been withdrawn and not replaced. General purpose board is the least expensive but can lose up to 60 per cent of its strength and increase by 10 per cent in thickness when wet. As with plywood the end use classification of particleboard depends on the bond strength of the adhesive used. Boards used for flooring or roof decking should be of a higher strength grade and where used in areas where dampness could affect the functioning of the board, moisture resistant boards using more durable adhesives or cement bonded particleboard should be used. A heavy-duty board is also available for areas where loading is likely to be higher than normal. Particleboard is generally used for floor and roof decking, but veneers can be attached to make it suitable for joinery applications.

Cement bonded particleboard (CBPB) comprises a mixture of timber particles, Portland cement, water and setting agents to *BS EN 634-2:2007 Cement-bonded particleboards. Specifications. Requirements for OPC bonded particleboards for use in dry, humid and external conditions*. The mixture is then place into a press and allowed to set with the aid of heat. Because the board has a high amount of Portland cement in its composition, it is very durable, has a high fire resistance and good acoustic insulation properties and has better dimensional stability than other particleboards. It is particularly useful for floor and roof decking in areas where there is a likelihood of water spillage or condensation occurring. It is also useful as a sheathing material in both timber frame and light steel frame house construction.

Oriented strand board (OSB) is similar in manufacture to particleboard, but similar in structure to plywood. It is covered by *BS EN 300:2006 Oriented strand boards (OSB). Definitions, classification and specifications*. It comprises strands of timber, up to 75mm in width, which are cut from logs and formed into a mat of fibres. The boards are generally three layers in thickness, with the orientation of the strands being changed in the outer two layers to that of those in the central layer. Synthetic adhesive is then sprayed onto the mats and they are placed in a press and heated to set the adhesive and form the board. Because of the alternating orientation of the strands in the three layers, OSB is stronger than ordinary particleboard and has enhanced dimensional stability. Its main applications are in floor or roof decking or as a sheathing material in timber frame and light steel frame house construction.

Flaxboard is manufactured in a similar manner to particleboard, but instead of using particles of timber it comprises chips from the stalk of the flax plant (flax is used in the manufacture of linen). These chips are mixed with synthetic resin binders and compressed to form boards. As with particleboard, the type of synthetic resin binder used determines the moisture resistance and hence the end use of the board. Flaxboard can be used for similar applications to those of particleboard and is covered by *BS EN 15197:2007 Wood-based panels. Flaxboards. Specifications*.

The sustainability credentials of particleboard and oriented strand board and flaxboard are quite high. Because of its method of manufacture, a high percentage of the timber used in particleboard can be obtained from recycled sources, and the timber used in both particleboard and OSB can be of low strength quality, which would not be able to be used for other structural applications. The flax used in flaxboard is a waste product from the linen manufacturing industry and therefore has high sustainability credentials due to its recycled material content. The high amount of cement used in the manufacture of cement-based particleboard reduces its sustainability credentials, but this is offset somewhat by the high durability of the product. The heat used in the manufacture of these boards increases their embodied energy content, but much of this heat is generated from the incineration of timber waste. The boards can be manufactured in this country, thus reducing the transportation energy requirements.

Fibreboard

These are produced by pulverising wood chips down to individual fibres by subjecting them to steam under pressure to soften them and then flaking them between two revolving discs. Two main types of fibreboard are produced, covered generally by *BS EN 622-1:2003 Fibreboards. Specifications. General requirements*. Medium density fibreboard (MDF) is manufactured by drying the fibres and mixing them with small quantities of a synthetic resin binder. The type of binder used determines the moisture resistance of the board. The resulting mixture is formed into mats and pressed into boards under heat and pressure. It is covered by *BS EN 622-5:2009 Fibreboard. Specification. Requirements for dry process boards (MDF)*. It has many applications, particularly in joinery and the manufacture of kitchen work surfaces. It is also used in flush and moulded door manufacture. High density and low density fibreboards do not use a resin binder, but instead the wet fibres are formed into boards by pressing them at high temperature into boards. The degree of compression determines the density of the board and to a great extent its strength.

Hardboard is a high density fibreboard and has a typical density of 900 kg/m^3. It is covered by *BS EN 622-2:2004 Fibreboard. Specification. Requirements for hardboards*. It does suffer from strength loss and swelling if it becomes wet. A tempered version can be used in damp situations. Hardboard can be used as a facing to internal flush doors, but higher quality doors generally use plywood facings. It is also used as a base layer in the manufacture of laminate flooring panels. Low density fibreboards are more commonly used as insulation boards and can also be used as wall or ceiling linings. They are covered by *BS EN 622-4:2009 Fibreboard. Specification. Requirements for softboards*. Because their strength is not of high importance some low density fibreboards can be manufactured using high amounts of recycled paper. The sustainability credentials of fibreboards are similar to those of particleboards previously considered.

 Visit the companion website to test your understanding of Chapter 10 with a multiple choice questionnaire, and to see 23 full-colour photos illustrating this chapter.

CHAPTER 11

Ground floors

Introduction

The primary function of all floors in a building is to provide support to the occupants of the building, along with any furniture and equipment. This chapter will consider the functional requirements of floors, both ground floors and upper floors, and compare and contrast how the functional requirements are affected by the location of the floor. The four main types of ground floor used in domestic construction are then considered.

Functional requirements

These relate to all floors, both ground floors and upper floors. However there are some distinct differences related to the location of the floor.

Strength and stability

The loads being carried by the floor need to be supported without causing excessive deflection (bending) of the floor and for the floor to be able to carry the loads to suitable supports. Ground floors will normally bear their loads directly onto the ground beneath, whereas upper floors will need to bear their loads onto supporting walls or beams. Where floors are fully supported by the ground beneath, deflection will be minimal and well within acceptable limits as long as the ground beneath is not subject to excessive seasonal movement, as may occur with shrinkable clay soils. However, when the floor is spanning freely between two supporting walls, the tendency for the floor to bend under the load will be increased. This deflection will be maximised at mid-span, the point furthest away from the supports. The depth of the supporting members of the floor (the joists) will need to be sufficient to reduce the deflection to within acceptable limits. This aspect is considered in more detail in the next chapter.

Fire resistance

Floors act as horizontal barriers to curb the spread of fire vertically throughout the building. This is normally only a requirement for upper floors.

Sound insulation

Floors need to reduce sound transmission between storeys. Again this is normally only a requirement for upper floors. It may be necessary to incorporate sound insulation materials within the construction of the upper floor.

Thermal insulation

Heat loss through the floor construction is normally only a problem with ground floors, where heat from the building may escape to the cooler area of the ground beneath. The extent of the heat flow through the floor construction will be dependent upon the thermal properties of the materials used in the floor construction, the temperature differential between the internal surface of the floor and the ground beneath, the climate, the physical properties of the soil, the presence and extent of ground water near to the surface, the amount of ventilation provided within the floor construction and whether or not underfloor heating is used. It will therefore be necessary to incorporate thermal insulation materials within the construction of the ground floor, and the detailing of the floor construction will need to ensure that thermal bridges are avoided and surface condensation and mould growth are also avoided. The Building Regulations Approved Document L1A Section 4.21 sets out limiting U-values for various elements of the building. These are weighted average values (as explained in Chapter 1) and for ground floors this value is $0.25W/m^2K$. In addition Approved Document C Section 4.22 stipulates that the ground floor should be designed and constructed so that the U-value does not exceed $0.7W/m^2K$ at any point and junctions between the floor and the walls are designed to avoid thermal bridging.

BS EN ISO 13370:2007 Thermal performance of buildings. Heat transfer via the ground. Calculation methods provides methods of calculation of heat transfer coefficients and heat flow rates for ground floors, both the annual average rate of heat flow and the seasonal variations of heat flow.

Damp penetration

Dampness may penetrate from the moist ground beneath the floor into the building by capillary action through the porous materials of the floor construction (refer back to the section on capillarity in Chapter 1 (introduction). Ground floors will therefore normally require a damp proof barrier to be incorporated within their construction to resist this rising damp.

The ground floor may also need to resist the passage of ground gases such as methane or radon (refer back to the section on contaminants in Chapter 3 (site investigation and the type of ground). In these circumstances a gas resistant barrier will need to be incorporated in the ground floor construction and this barrier may also function as a damp proof membrane.

Ground floors may be of either solid or suspended construction. Solid construction is generally the easiest and cheapest to construct.

Solid floor construction

The floor is constructed on a bed of well-compacted hard material, called hardcore. It provides the following benefits:

- a level, horizontal surface on which to place the floor slab
- a firm, dry working surface on which to place the concrete for the floor slab
- it reduces the rise of ground moisture by virtue of the large voids between the pieces of hardcore which eliminate capillary paths
- it provides additional support to the floor above should the ground beneath suffer settlement or subsidence.

Self-assessment question 11.1

a) Why should the floor slab require a level, horizontal surface on which to be laid?
b) Why should the floor slab require a firm, dry surface on which to be placed?

Hardcore consists of broken bricks, stones, concrete or rubble. It must not contain materials that are soft or can crumble easily and must not contain deleterious material that will either rot or corrode, leaving voids, or contain sulphate salts that could react with the concrete.

Self-assessment question 11.2

Why is it important that hardcore should not contain materials that are soft or can crumble easily?

The lumps of hardcore should be relatively large, but generally not larger than 75mm in diameter.

Self-assessment question 11.3

Why is it important that the lumps of hardcore should be relatively large?

It should be laid to a minimum thickness of 100mm, be well-compacted and have its surface 'blinded' with a layer of sand. This helps to fill in the large voids between the pieces of hardcore on the surface, preventing loss of concrete into these voids, and also covers the sharp arrises on the surface pieces of hardcore that could puncture a damp proof membrane placed above.

The damp proof membrane (DPM) prevents the rise of ground moisture through the floor construction. In order for it to be functional it must be continuous, impervious and join up with the damp proof courses in the walls.

Self-assessment question 11.4

Why should the damp proof membrane join up with the damp proof course in the walls?

There are three positions in which the damp proof membrane may be located within the floor construction:

- sandwiched between the blinding above the hardcore and the concrete slab
- sandwiched between the top of the concrete slab and the screed
- sandwiched between the surface of the slab and the floor finish.

If the DPM is to be placed between the blinding and the concrete slab it will normally comprise a 1200 gauge low density polyethylene sheet to *BS EN 13967:2004 + Amendment 1:2006 Flexible sheets for waterproofing. Plastic and rubber damp proof sheets including plastic and rubber basement tanking sheet. Definitions and characteristics* (see Figure 11.1). This sheet, being of finite width, must be sealed effectively at its edge joints. Alternatively bitumen sheet to *BS EN 13969:2004 Flexible sheets for waterproofing. Bitumen damp proof sheets including bitumen basement tanking sheet. Definitions and characteristics* may be used as a damp proof membrane in this position.

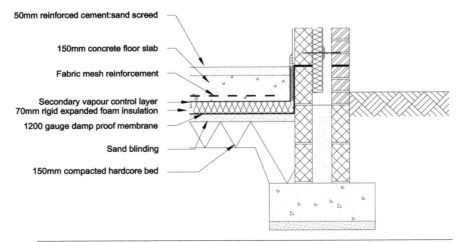

50mm reinforced cement:sand screed
150mm concrete floor slab
Fabric mesh reinforcement
Secondary vapour control layer
70mm rigid expanded foam insulation
1200 gauge damp proof membrane
Sand blinding
150mm compacted hardcore bed

Figure 11.1 Solid ground floor with DPM below the floor slab

Self-assessment question 11.5

Why does the polyethylene sheet need to be sealed effectively at its edge joints?

If the DPM is to be placed above the concrete slab then it may comprise a 1200 gauge polyethylene sheet or bitumen sheet as previously considered, or 3 coats of a cold applied bituminous solution that can be applied by brush, or 2 coats of hot applied mastic asphalt similar to the tanking used in basements. In these cases the screed will need to be thicker and reinforced with chicken wire to prevent cracking, as it will not be bonded to the slab beneath (see Figure 11.2). This will be considered in more detail in Chapter 21 (internal finishes).

Self-assessment question 11.6

What disadvantage does the cold applied bituminous solution have compared with the 1200 gauge polyethylene sheet as a DPM?

Where the DPM is applied to the surface of the floor slab it should comprise 3 coats of a cold applied bituminous solution that can be applied by brush, or 2 coats of hot applied mastic asphalt as considered previously.

The concrete slab should be a minimum of 100mm thick to mix ST2 in *BS 8500-1:2006 Concrete*, and the surface either finished with a float or tamped to receive a screed for an applied finish. Tamping involves bouncing a timber beam over the surface of the concrete slab. The beam is normally operated by an operative at each end and it produces a ripple effect to the surface of the concrete slab.

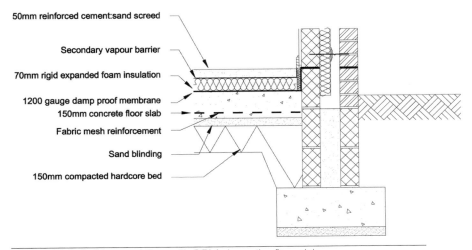

50mm reinforced cement:sand screed

Secondary vapour barrier

70mm rigid expanded foam insulation

1200 gauge damp proof membrane
150mm concrete floor slab

Fabric mesh reinforcement

Sand blinding

150mm compacted hardcore bed

Figure 11.2 Solid ground floor with DPM above the floor slab

Self-assessment question 11.7

What benefits does tamping provide to the concrete slab?

The slab may also contain a layer of fabric mesh reinforcement, which helps to provide some structural strength to the slab over areas of potential weakness. The concrete to a reinforced slab should be to mix ST4 in *BS 8500*.

In order to comply with the requirements of the Building Regulations, Approved Document L1A 2010 Conservation of fuel and power, stipulates that thermal insulation must be incorporated into the ground floor construction to prevent loss of heat from the building to the ground beneath. Section 4.21 of Approved Document L1A sets out in Table 2 the worst acceptable U-value standards for various elements of the building fabric to meet the energy performance requirements of a new dwelling. The U-values are calculated using the methods and conventions set out in BR 443 Conventions for U-value calculations 2006. For a floor this maximum U-value is $0.25W/m^2K$.

The insulation should, where possible, be laid above the DPM and should also be placed at the side of the slab where the slab abuts the walls, so that heat will not be lost from the building through this path. The insulation will normally be a board material to prevent compression by the weight of the floor slab or the occupants, furniture and equipment above.

Self-assessment question 11.8

Why is it important that the insulation is, where possible, laid above the DPM?

If the insulation layer has to be placed below the DPM it should comprise a material that has low water absorption and is resistant to contaminants in the ground, where they exist.

Suspended floor construction

Suspended floors do not bear directly onto the ground but bear onto supports that will themselves indirectly bear onto the ground. They are often utilised on sloping sites to reduce the amount of fill materials. Fill will be required beneath the ground floor on sloping sites in order to provide a level, horizontal surface on which to lay the floor. It will normally comprise hardcore not greater than 600mm in thickness.

Self-assessment question 11.9

Why should the hardcore fill not be greater than 600mm in thickness?

Suspended floors may be constructed from timber or concrete.

Suspended timber ground floor

As timber is warmer to the touch than concrete, this floor may be considered to provide increased thermal comfort to the user.

The floor comprises small timber beams or joists which span between points of support called dwarf walls and carry floor decking of timber boards or particle board or plywood sheets.

The joists are sized according to the loads that are to be carried and the distances that are to be spanned.

Suitable sizes are given in the Building Regulations Approved Document A. However, a rule of thumb method may be used:

$$\text{Depth in mm} = \frac{\text{Span in mm}}{24} + 50\text{mm}$$

Self-assessment question 11.10

Using the rule of thumb method, what would be the depth of a joist spanning 2.00m between supports?

The floor is constructed on a 100mm thick concrete oversite slab to mix ST1 in *BS 8500-1*, laid on a minimum 100mm thick layer of compacted hardcore. The Building Regulations Approved Document C section 4.14 stipulates that the top of the slab must not be below ground level, or on sloping sites land drainage needs to be installed to the outside of the building on the upper slope. This is to prevent water seeping from the surrounding ground and collecting on the top of the slab.

The joists are supported on half brick dwarf or sleeper walls, spaced at 1–2m centres and topped by a 100 x 75mm wall plate, laid over a damp proof course.

The wall plate is used to provide a suitable fixing for the floor joists, as it is difficult to fix the joists directly to the sleeper walls. The wall plate itself is fixed to the sleeper walls by means of galvanised steel straps (see Figure 11.3).

Timber is susceptible to attack by dry rot fungus if its moisture content is allowed to increase above 20 per cent of its dry weight (see Chapter 10 on timber). The air in the void beneath the timber members in the floor will have an artificially high moisture content, caused by the rising dampness from the ground beneath.

A DPM should be incorporated between the concrete oversite slab and the hardcore. However in areas such as kitchens, utility rooms and

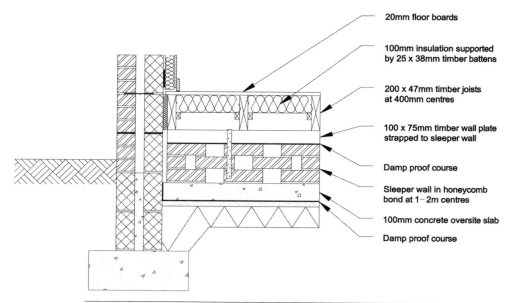

20mm floor boards

100mm insulation supported by 25 x 38mm timber battens

200 x 47mm timber joists at 400mm centres

100 x 75mm timber wall plate strapped to sleeper wall

Damp proof course

Sleeper wall in honeycomb bond at 1–2m centres

100mm concrete oversite slab

Damp proof course

Figure 11.3 Suspended timber ground floor

bathrooms, where water may be spilled and could remain undetected on the top of the oversite concrete slab and would not have the opportunity to drain away, any board used as flooring should be moisture resistant.

Self-assessment question 11.11

Why should a damp proof course be laid beneath the wall plate?

The Building Regulations Approved Document C Section 4.14 requires a ventilated air space from the ground covering to the underside of the wall plate of at least 75mm and a minimum space to the underside of the joists of 150mm, which must be kept free from debris. Where suspended timber floors are constructed in areas where shrinkable clay subsoils are prevalent, the depth of the underfloor air space may need to be increased to allow for any subsequent ground heave.

In areas where flooding may occur, a means of inspecting and cleaning out the space beneath the suspended floor should be provided.

Ventilation to the sub-floor space is provided by air bricks within the external walls, which should incorporate suitable grilles to prevent the entry of vermin into the underfloor area. These air bricks must provide a minimum opening of 1500mm^2 per metre run of wall, or 500mm^2 per square metre of floor area, whichever is the least. This ventilation helps to remove the damp air beneath the floor construction and replace it with drier air from outside. Unfortunately this replacement air is not only drier but also colder. In order to ensure that this air flow does not detrimentally affect the thermal insulation value of the floor it is recommended that the amount of openings in the external wall should be sufficient to encourage

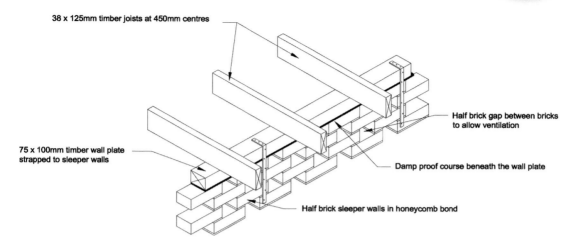

38 x 125mm timber joists at 450mm centres

Half brick gap between bricks
to allow ventilation

75 x 100mm timber wall plate
strapped to sleeper walls

Damp proof course beneath the wall plate

Half brick sleeper walls in honeycomb bond

Figure 11.4 Honeycomb bonding to sleeper walls

cross flow ventilation, but not so much that the thermal insulation value
of the floor will be detrimentally affected.

The cross flow of air within the void beneath the timber floor is
maintained by building the sleeper walls in a honeycomb bond, which
consists of a stretcher bond with voids of half brick width being left
between the stretcher bricks (see Figure 11.4).

Self-assessment question 11.12

Why should the space to the underside of the joists be kept free from
debris?

Thermal insulation can be in the form of quilt draped over the top of
the joists or a slab spanning between the joists and supported by timber
battens nailed to the side of the joists.

The boarding to a suspended timber ground floor should be from
durable species or treated with preservative and should be a minimum of
20mm in thickness. Alternatively moisture resistant grade plywood,
particleboard, oriented strand board or flaxboard may be used. The gap
between the edge of the boarding and the skirting around the base of the
walls must be caulked to provide a complete seal from draughts.

Suspended concrete ground floor

This form of construction overcomes three problems:

- dry rot in timber floors
- excessive fill beneath floors of buildings situated on steeply sloping
  sites
- variable loadbearing capacity soil beneath the floor.

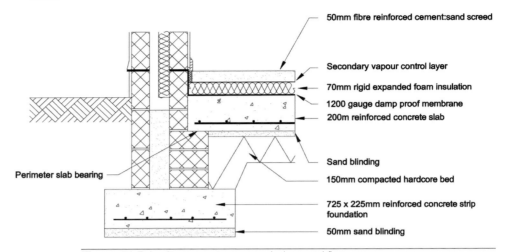

50mm fibre reinforced cement:sand screed

Secondary vapour control layer

70mm rigid expanded foam insulation

1200 gauge damp proof membrane

200m reinforced concrete slab

Sand blinding

150mm compacted hardcore bed

725 x 225mm reinforced concrete strip foundation

50mm sand blinding

Perimeter slab bearing

Figure 11.5 In situ concrete suspended ground floor

The floor may be constructed from in situ concrete, similar to a solid floor or from precast concrete components.

Where the depth of fill will exceed 600mm on steeply sloping sites or where the loadbearing capacity or nature of the ground beneath the floor varies, a floor bearing directly onto the ground may not be considered suitable. In such cases the floor will need to span between loadbearing walls.

An in situ concrete suspended ground floor slab should be at least 100mm in thickness, containing at least 300kg of cement per cubic metre of concrete and incorporating sufficient steel reinforcement in its lower section to enable it to effectively carry the loads bearing upon it between the loadbearing wall supports, protected by a concrete cover of at least 40mm (see Figure 11.5).

As the floor is cast, the concrete will require support until it has hardened sufficiently to carry its own weight. It will therefore be necessary to place fill materials onto the ground to provide a firm, dry, level horizontal surface onto which the slab may be cast. Although this fill will need to be compacted initially to provide a firm surface for the slab, any subsequent settlement of this fill will not adversely affect the floor above, since the slab will be supported by the loadbearing walls.

Self-assessment question 11.13

Where should the DPM be placed in an in situ concrete suspended ground floor?

The precast concrete suspended ground floor utilises precast concrete joists, infilled with concrete blocks and topped with a screed. The joists and blocks are light enough to be lifted and positioned by two men, thus eliminating the need for lifting plant.

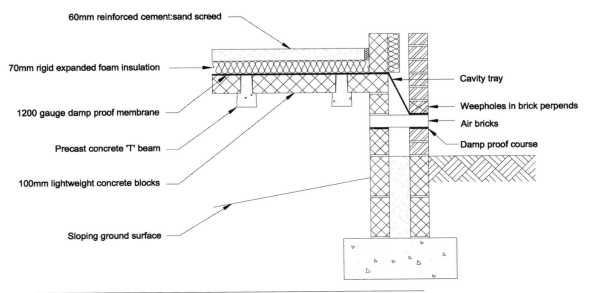

60mm reinforced cement:sand screed

70mm rigid expanded foam insulation

1200 gauge damp proof membrane

Precast concrete 'T' beam

100mm lightweight concrete blocks

Sloping ground surface

Cavity tray

Weepholes in brick perpends

Air bricks

Damp proof course

Figure 11.6 Precast concrete suspended floor

The joists are generally an inverted T shape and are prestressed to reduce their size and weight. They span between external walls and, as the precast concrete components do not require support during the construction of the floor, the need for fill beneath the floor is eliminated. Thus all that will be required is for the topsoil to be removed from the ground beneath the floor and for the subsoil to be treated with a weedkiller prior to the installation of the floor components.

Thermal insulation is generally placed above the precast concrete beam and block components and beneath the screed (see Figure 11.6).

A minimum void of 150mm should be provided between the soil beneath and the underside of the floor slab in an in situ concrete suspended floor or the underside of the precast concrete rib beams in a precast concrete suspended floor. As before, this space may need to be increased to account for any possible heave where the soil beneath is shrinkable clay. The underfloor void created should be ventilated by airbricks positioned in the external walls to create 1500mm^2 of ventilation per metre run of wall or 500mm^2 per square metre of floor area as before.

This ventilation is also beneficial where gas is supplied to the property, since any potential gas leaks could collect in the void beneath an unventilated suspended floor and build up to form an explosive mixture with the air.

Visit the companion website to test your understanding of Chapter 11 with a multiple choice questionnaire, and to see 16 full-colour photos illustrating this chapter.

CHAPTER **12**

Upper floors

Introduction

Upper floor construction in houses is based on similar principles to that of suspended timber ground floor construction, the difference being that the spans between loadbearing supports are generally greater with upper floor construction, so the joists need to be larger in size. Although in situ concrete floors and precast beam and block floors could be used for upper floor construction, they are not popular and suspended timber construction is still the most widely used method.

This chapter will consider the construction of upper floors including the methods of support, the use and construction of double floors, framing of upper floors around openings, joints used in upper floor construction, silent floor construction, floor decking and strutting to joists.

Self-assessment question 12.1

Why will the spans between loadbearing supports be greater in upper floor construction than they are in suspended timber ground floor construction?

Sizing of joists

The sizes of joists in upper floor construction are determined by the load being carried, the span of the joist and the spacing of the joist. Section sizes and spans for floor joists are now provided in Eurocode 5 span tables: for solid timber members in floors, ceilings and roofs for dwellings, published by the Timber Research and Development Association, which complies with *BS 5268-2:2002 Code of practice for permissible stress design, materials and workmanship.* From these tables it is possible to ascertain that by choosing timber with a higher strength grade, it is possible to span greater distances without having to increase the size of the joist.

Joists to upper floors are normally spaced at 400 or 600mm centres. This is because the plasterboard sheets that are mainly used for the ceiling finish come in modular sizes (normally 2400mm long x 1200mm wide) and to enable satisfactory fixing of the plasterboard, the edge of abutting sheets should coincide with the centre of a joist.

Self-assessment question 12.2

Why should the spacing of the joists have an effect on their size?

Support

The ends of the joist will need to be supported by loadbearing walls. Where the support is provided by the external cavity walls, the end of the joist will bear onto the internal leaf. The Building Regulations Approved Document A, Section 2C23 stipulates that the maximum span for a floor supported by a wall should be 6m. Because the span from one external wall to the other may therefore be too excessive, it is normal practice to utilise loadbearing internal walls to also support the upper floor joists. Joists may either bear directly onto the wall, in which case they will be built in to the external wall (see Figure 12.1) and rest on top of the ground to first floor internal wall (see Figure 12.2), or they may be supported on joist hangers (see Figure 12.3).

Self-assessment question 12.3

What are the advantages and disadvantages of using joist hangers to support the upper floor joists?

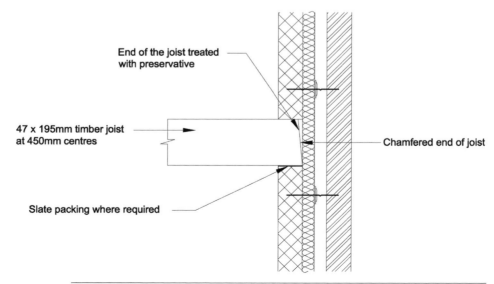

End of the joist treated with preservative

47 x 195mm timber joist at 450mm centres

Chamfered end of joist

Slate packing where required

Figure 12.1 Support to the joist at an external wall

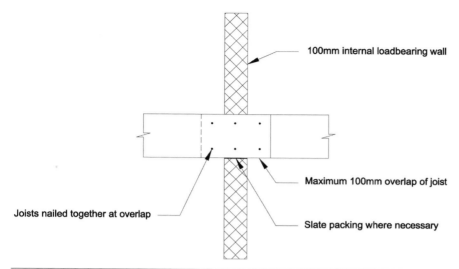

Figure 12.2 Support to the joist at an internal wall

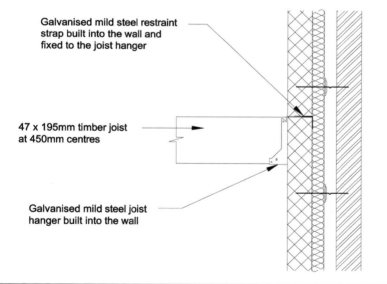

Figure 12.3 Support to the joist by the use of a joist hanger

Where the joist bears onto the wall, it may be necessary to provide packing to the underside of the joist so that the top surfaces of the joists are level. This may be required because the sizes of unplaned timbers may differ slightly. Pieces of slate have traditionally been used for this packing, but any thin material that will not decompose over time and will be capable of bearing the load imposed upon it by the joist without breaking up may be used.

Why should the tops of the joists need to be level?

Building Regulations Approved Document E, Section 2.45 also stipulates that in order to avoid flanking sound transmission, where a floor is supported on a solid masonry separating wall, then the joists should be supported on hangers and not built in.

Also, to avoid thermal bridging between the floor and the supporting external wall, thermal insulation needs to be provided at the junction between the upper floor and the external wall.

Where the joists are to be built in to the supporting wall their ends need to be treated with preservative. It is also a good idea, where the end of the joist is facing the cavity in an external wall, to chamfer the end so that any water that may drip onto the end of the joist from the wall ties above can be drained off quickly.

Where the joists are built in to internal loadbearing walls their ends should overlap by not more than 100mm each side of the wall and the ends of the joists should be nailed together. Alternatively, they should be supported on joist hangers.

The upper floors not only provide support for additional accommodation within the building, they also provide lateral stiffening to the loadbearing walls to which they are attached. In order for the floors to fulfil this additional function effectively, the joists need to be strapped to the wall with galvanised steel or austenitic steel tension straps conforming to *BS EN 845-1:2003 + Amendment 1:2008 Specification for ancillary components for masonry. Ties, tension straps, hangers and brackets*.

The Building Regulations Approved Document A, Section 2C35 states that in houses of not more than two storeys and where the joists are placed at less than 1.2m centres and have a bearing of at least 90mm on the supporting walls or use restraint type joist hangers to *BS EN 845-1*, then they do not require tension straps to secure them to the walls in the longitudinal direction. However, where the joists run parallel to an external wall, restraint straps should be placed at a maximum spacing of 2m and fixed over 3 adjoining joists, supported by *noggings* (see Figure 12.4).

Why should the external walls need to be stiffened at upper floor level?

Double floors

Occasionally the span of the floor between wall supports is too great and an intermediate support is required that will not interfere with the use of

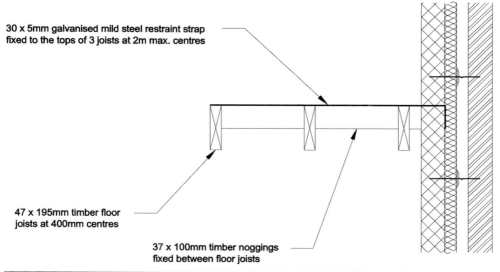

30 x 5mm galvanised mild steel restraint strap
fixed to the tops of 3 joists at 2m max. centres

47 x 195mm timber floor
joists at 400mm centres

37 x 100mm timber noggings
fixed between floor joists

Figure 12.4 Restraint straps to joists running parallel to an external wall

the room beneath the upper floor. This intermediate support could be a timber beam that the ends of the joists could bear on or, more commonly, a *rolled steel joist* (RSJ). The RSJ spans between two loadbearing walls and the ends of the joists are notched in to the RSJ, allowing room for the decking to be fixed to the upper surface of the joists, the ceiling finish to be fixed to the lower surface of the joists and a space allowed for timber shrinkage. This form of floor is called a double floor (see Figure 12.5).

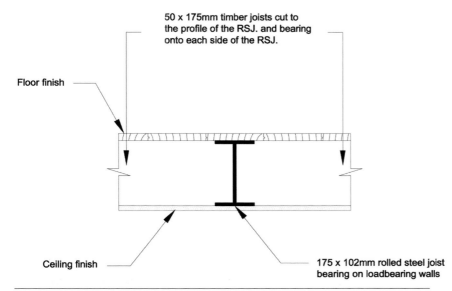

50 x 175mm timber joists cut to
the profile of the RSJ. and bearing
onto each side of the RSJ.

Floor finish

Ceiling finish

175 x 102mm rolled steel joist
bearing on loadbearing walls

Figure 12.5 Double floor construction

Framing around openings

Openings must be provided in upper floors to allow for stair access and the passage of chimney flues. Building Regulations Approved Document A, Section 2C37 stipulates that where an opening for a stairway adjoins a supported wall and interrupts the continuity of lateral support provided by the upper floor to the wall, then the maximum permitted length of the opening should be 3m, measured parallel to the supported wall. Also, where a connection is provided by mild steel anchors, these should be spaced closer than 2m on each side of the opening to provide the same number of anchors as if there were no opening. These openings need to be formed by framing the upper floor joists around the opening (see Figure 12.6).

There are four types of joist used for this framing:

- the common joist – this is a full length joist spanning between wall or RSJ supports without interruption.
- the trimmed joist – this is similar to the common joist but has been cut short to form the opening. It spans in the same direction as the common joist and one end is supported by a wall or RSJ and the other end is supported by the trimmer joist.
- the trimmer joist – this joist forms the side edge of the opening and supports the cut ends of the trimmed joists. It spans at right angles to the common and trimmed joists.
- the trimming joist – this forms the front and rear edges of the opening and supports the ends of the trimmer joist. It spans in the same

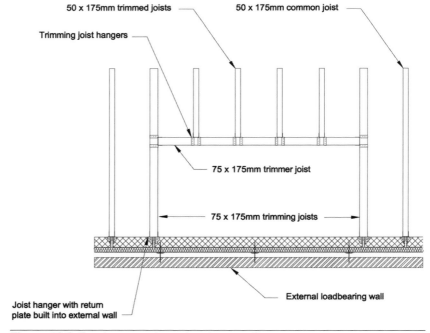

Figure 12.6 Framing around openings in upper floors

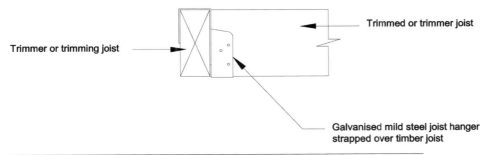

Figure 12.7 An adapted joist hanger supporting joists around the formation of an opening

direction as the common and trimmed joists and, similar to the common joist, it spans between wall or RSJ supports without interruption.

Because the trimmer and trimming joists are required to carry a greater load than the common and trimmed joists, they are generally 25mm thicker. However, it is now common to see two regular thickness joists nailed together to form trimmer or trimming joists.

Self-assessment question 12.6

Why do trimmer and trimming joists need to be made wider rather than deeper than common or trimmed joists to increase their strength?

Traditionally in forming this framing to openings, various carpentry joints were used to connect the joists. However, it is now common practice to see adapted joist hangers being used for this purpose. The joist hanger may be strapped to the supporting joist and used to support the cut end of the supported joist (see Figure 12.7).

Silent floors

The biggest problem with using solid timber for upper floor joists is that it can suffer from moisture movement that can lead to distortion of the floor or, more commonly, squeaking of the floor boards or decking material when a load is applied to the floor, caused by some of the supporting joists having shrunk away from the floor boards or decking above.

This problem can be overcome by using prefabricated timber 'I' section joists, manufactured from laminated timber *flanges* with a plywood *web* in place of the solid timber joists (see Figure 12.8). Laminated timber is a method of being able to build up timber sections by gluing together thin strips of timber or *laminas* to produce a compound section that will have similar structural properties to a solid

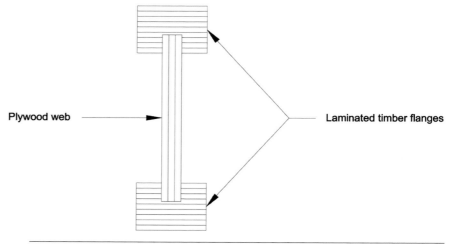

Figure 12.8 A silent joist

timber section of the same size. The advantage of using laminated timber is that large section timber members can be made at a fraction of the cost of the equivalent solid timber section, the timber used for the laminas is carefully selected and is defect free and the section produced suffers less moisture movement because it is prefabricated.

A similar system is available utilising strength graded timber flanges joined together by V-shaped galvanised steel web sections which are fixed to the flanges by nail plates (see Figure 12.9). These have the advantage of an open web system giving space for the installation of

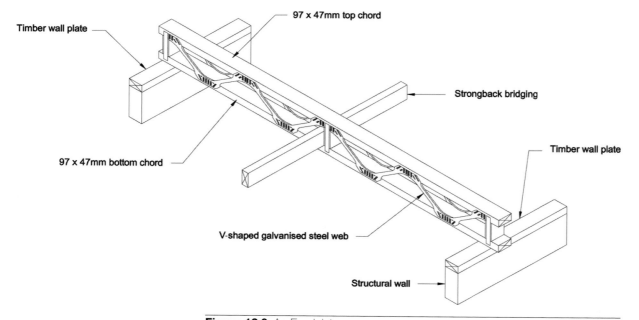

Figure 12.9 An Eco joist

services without cutting the joist. They are lightweight thus saving time in erecting them.

The 'silent' joists produced by these techniques are of a similar section size and strength to solid timber joists, but weigh considerably less and suffer less drying shrinkage, thus virtually eliminating the squeak of floor boards and decking. They are however more expensive than solid joists.

Floor decking

Decking is fixed to the upper surface of the floor joists to provide a suitable surface to receive applied floor finishes. It may comprise:

- softwood timber boards – 125mm wide x 16mm thick (for 400–450mm joist spacing) or 19mm thick (for 600mm joist spacing) with tongued and grooved joints;
- plywood sheets – 2.4 x 1.2 m sheets, 12mm thick (for 400–450mm joist spacing) or 16mm thick (for 600mm joist spacing) with tongued and grooved joints;
- particleboard sheets – 2.4 x 1.2m sheets, 18mm thick (for 400–450mm joist spacing) or 22mm thick (for 600mm joist spacing) with tongued and grooved joints. Because particleboard sheets can swell and lose inherent strength when wet, it is recommended that where the floor decking is to be laid in bathrooms or other areas where the decking may be susceptible to water spillage, a cement bonded particleboard be specified. This board has high moisture resistance and also has high fire resistance properties. Alternatively, sheets of oriented strand board or flaxboard may also be used, with particular attention being given to the resin used in their manufacture.

Tongued and grooved joints are used in floor decking for the following reasons:

- They help to spread the load on the decking. Without side joints between the decking materials any applied load would need to be carried by the board or section of decking that it was bearing on. Part of this load could not be transferred to adjacent boards.
- They improve thermal insulation by reducing air infiltration or draughts through the gaps between adjacent boards. This is more important in ground floor construction than it is in upper floor construction.
- They improve the fire resistance of the decking by reducing the infiltration of smoke and flames through the gaps between adjacent boards.

Sheets of plywood, particleboard and similar materials have become more popular than softwood boards as floor decking materials, since they can be laid much more quickly. However, because the void between the floor decking and the ceiling finish, which is created by the depth of the floor joists, is so useful for the running of electrical cable and plumbing pipes throughout the house, it may be necessary, occasionally, to gain access to this void for maintenance purposes. It is far easier to take up a

125mm softwood floorboard to gain access to this underfloor space than it is a 2.4 x 1.2m sheet of plywood, particleboard or similar materials. Where these materials are to be used for the floor decking, installers should provide suitable access panels that can easily be removed and replaced.

Where pipework and cables are installed in the underfloor space they should run through holes cut in the middle of the joists, around the neutral axis, where the timber removed will have the least effect on the structural strength of the joist.

The resistance to airborne sound transmission through the upper floor depends on the structural floor base, the construction of the ceiling and the use of absorbent material in the ceiling void. However, this is generally only a problem where the floor is regarded as a separating floor between flats in a building of multiple occupancy. In a house where the upper and ground storeys are in the same occupancy, the most important consideration regarding sound transmission through the upper floor is that of impact sound. This can be reduced considerably by applying a soft covering as a finish to the upper floor and the inclusion of a 100mm layer of mineral wool insulation in the floor void.

Strutting to joists

As previously discussed, solid timber joists may swell or shrink with changes in their moisture content. If joists are restrained from longitudinal movement at fixings then any shrinkage or swelling will cause twisting to occur within the joist member. This will cause cracking in the floor decking above and/or the ceiling finish below. Restraints in the form of strutting should be provided between joists to prevent twisting occurring. The strutting is not necessary where the floor is of short span (less than 2.5m) but should be provided at mid-span in floors spanning between 2.5 and 4.5m and at one-third span in floors spanning over 4.5m.

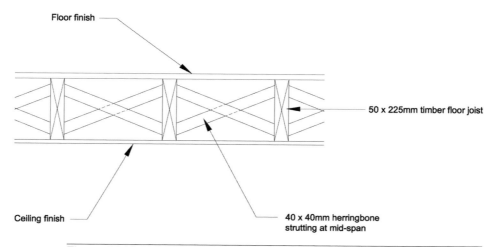

Floor finish

50 x 225mm timber floor joist

Ceiling finish

40 x 40mm herringbone strutting at mid-span

Figure 12.10 Herringbone strutting

The strutting may be pieces of solid timber of similar size to the joists or can be 40 x 40mm diagonal braces called *herringbone strutting* (see Figure 12.10). A version of this herringbone strutting can also be created with the use of pressed steel diagonal braces.

Visit the companion website to test your understanding of Chapter 12 with a multiple choice questionnaire, and to see 30 full-colour photos illustrating this chapter.

CHAPTER **13**

Flat roof construction and coverings

Introduction

Flat roofs are the simplest form of roof construction and the least expensive to construct. However they can also be the most troublesome, since they have a tendency to leak and need much more regular maintenance than pitched roofs. There are a number of reasons for this and these will be considered in this chapter. Because of the problems associated with flat roof construction this form of construction tends to be avoided where possible. However, there are situations where a flat roof may be considered to be a more acceptable option than a pitched roof, for instance on an extension where there is not the room to accommodate the height of a pitched roof or where a pitched roof may look out of keeping with either the building or its surroundings.

In this chapter the functional requirements of roofs, both pitched and flat, are considered. The two main forms of flat roof construction, hollow suspended timber construction and solid suspended concrete slab construction are then discussed and compared. One particular problem with roofs is that of interstitial condensation. The causes of this are discussed and suitable remedies are considered.

Coverings to flat roofs need to be impervious, durable and have low resistance to the surface spread of flame. There are three main types of covering used for domestic flat roof construction and the properties and performance of each type of covering will be considered, along with their fixings at the eaves, the verges and the abutment. The problem of solar radiation and its effects on flat roof coverings and how this problem may be alleviated will also be considered.

The functional requirements of a roof

The primary function of a roof is to enclose space and protect it from the elements. This is similar to the primary function of an external wall. In fact roofs are sometimes referred to as 'the fifth wall' because they have similar functional requirements.

The main functional requirements of a roof are:

- strength and stability
- weather resistance
- thermal insulation
- sound insulation
- fire resistance
- durability
- appearance.

The dissimilarity between roofs and walls with regard to the functional requirements of these elements is that whereas walls perform all functions with one structure, roofs can be separated into two distinct sub-elements, that of the structure and the finish. The structure performs the functional requirement of strength and stability, fire resistance and, to a certain degree, appearance. The covering performs the functional requirements of weather resistance, durability and also contributes to the appearance. Thermal insulation and sound insulation have to be provided by other materials that are incorporated into the construction of the roof.

In order to overcome a certain amount of repetition in this chapter and the succeeding next chapter on pitched roof construction and finishes, the functional requirements of roofs, both flat and pitched, and their associated coverings will be considered here.

Strength and stability

The span of the roof has a major influence on its design. The materials used in roof construction need to have a high strength:weight ratio. Roofs used in domestic construction use members that span fairly small distances and which are located at relatively close centres. Timber is used to good effect for the structural framework but concrete can also be used in flat roof construction. The loads carried by roofs are mostly dead loads, but superimposed loads such as wind loading and snow loading need also to be considered. Occasionally flat roofs may have foot traffic for maintenance, or the flat roof of a projecting ground storey room or extension could be used as a balcony to a room on the first storey.

The effect of wind pressure is a major consideration in roof design. The wind blowing onto a building will create positive and negative pressures at certain points and will also create eddies as the wind is deflected by the shape of the building (see Figures 13.1 and 13.2).

Roofs of low pitch are subjected to high suction forces, particularly around *verges, eaves, the ridge* and *chimneys* (see Figure 13.3).

It is therefore important to anchor the roof structure to the supporting walls and also to secure the coverings to the roof structure and protect them against the action of the wind. Building Regulations Approved Document A, Section 2C36 stipulates that there should be a wall plate at the eaves of a flat roof. The joists should be connected to this with either framing anchors or skew nailing. The wall plate should be anchored to the wall with vertical straps at least 1m in length and at no greater than 2m

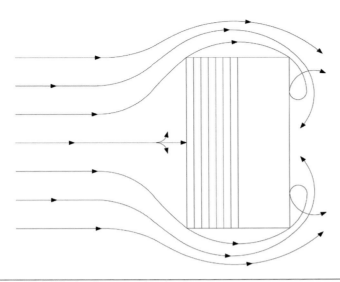

Figure 13.1 Wind paths around a structure

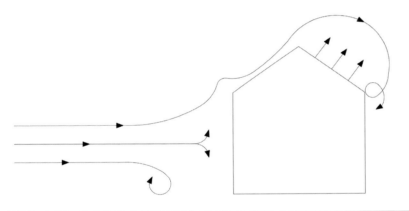

Figure 13.2 Wind paths over a roof

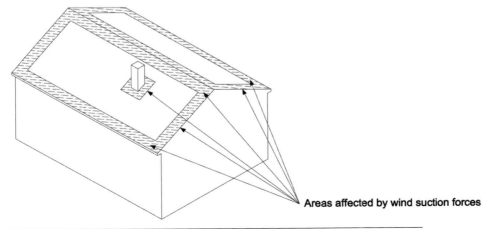

Areas affected by wind suction forces

Figure 13.3 Areas on a roof affected by high suction forces from the wind

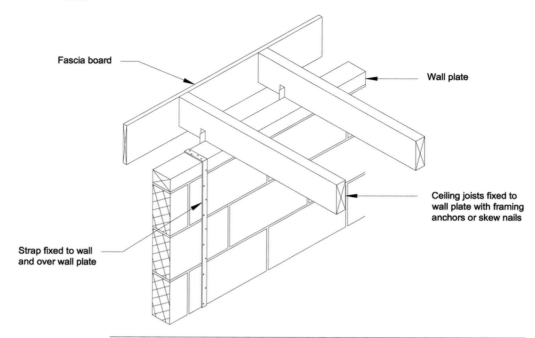

Fascia board

Wall plate

Ceiling joists fixed to
wall plate with framing
anchors or skew nails

Strap fixed to wall
and over wall plate

Figure 13.4 Vertical strapping of a wall plate at the eaves of a flat roof

centres (see Figure 13.4). For pitched roofs the vertical strap at eaves level should be attached to the rafter and the wall. In addition there should be tension straps attached to the gable wall and fixed to three rafters and timber noggins between the rafters at no greater than 2m centres.

Weather resistance

The covering and the provision of a suitable slope for the run-off of rainwater provide the majority of weather resistance for roofs. Water penetration through the covering is prevented either by the use of impermeable coverings on flat roofs or the use of overlaps between tiling units on pitched roofs.

Thermal insulation

The roof area constitutes a large part of the exposed surface area of a building, particularly if it is a hipped roof on a bungalow. Warm air rises; therefore heat losses through the roof are quite significant. Although most roofs on domestic buildings are constructed from timber, the amount of timber contained within the roof construction is relatively small and therefore the materials used in the roof construction contribute little to the thermal insulation value of the roof. Thus it is necessary to add other materials into the roof construction in order to provide adequate thermal insulation to the roof.

As with external walls and ground floors, the Building Regulations Approved Document L1A Section 4.21 sets out limiting U-values for various elements of the building. These are weighted average values (as explained in Chapter 1) and for roofs this value is 0.20W/m²K.

A particular concern in roof construction is the problem of *interstitial condensation*. The Building Regulations Approved Document C, Section 6.14 stipulates that the roof should be designed and constructed such that its U-value does not exceed 0.35W/m²K at any point in order to provide resistance to the development of surface condensation and mould growth. Techniques to combat this and provide an effective level of thermal insulation have led to the development of three main forms of design in flat roofs. These will be considered in more detail in the later section on flat roof coverings.

Sound insulation

Sound insulation is not as great a problem in roof construction as it is in wall construction. This is because the greatest source of external noise in houses is that emanating from traffic at ground level. The intensity of sound decreases considerably with the distance covered, so although the roof is of relatively lightweight construction compared with the walls, there is very little noise from traffic entering the house through this route. However, if overhead noise, particularly from aircraft, is a problem, then sound insulation materials will need to be added to the roof construction.

Fire resistance

The main consideration here is to protect the building from the spread of fire from adjacent buildings. Some materials, particularly on flat roofs, are flammable and may support the surface spread of flame. They must therefore be covered with a non-flammable material to provide adequate fire resistance to the roof.

The roof structure must also prevent the early collapse of the roof in a fire, as this would enable more oxygen to be available to the fire causing it to blaze more fiercely. It would also inhibit evacuation from the building during a fire and make the task of the fire fighting services in extinguishing the fire more difficult.

As will be discussed further in Chapter 20 (internal walls and partitions), where a separating party wall is provided between semi-detached or terraced houses, the wall will need to be taken up to the underside of the roof covering or decking and fire stopped where necessary. Also, to prevent a fire spreading over the roof from one house to an adjoining house, the separating party wall should either be extended through the roof for a height of 375mm above the surface of the roof covering, or there should be a zone of the roof extending to 1500mm on either side of the separating party wall in which the roof covering should have a low combustibility rating.

Durability

The durability of the roof is influenced to a great extent by the performance of the roof covering. Different materials can be used for roof coverings and each material has its own response to frost attack, chemical attack and solar radiation. These features will be considered in more detail in the sections on flat roof coverings in this chapter and pitched roof coverings in the next chapter.

The durability of the structure of the roof is mostly affected by the durability of the timber used in the roof construction. This aspect was considered in detail in Chapter 10 (timber).

Appearance

This is influenced by the type of roof, whether it is flat or pitched, the style of the roof, whether it is monopitch, dual pitch or mansard, and the materials used for the covering, whether they are sheet or tiles and if tiles, what shape, colour and texture.

Hollow suspended timber construction

This roof comprises timber joists spanning between supporting walls and carrying a deck onto which is laid the covering. This form of construction is very similar to that used for timber suspended upper floor construction considered in the previous chapter. As with joists for upper floors, section sizes and spans for roof joists are now provided in Eurocode 5 span tables for solid timber members in floors, ceilings and roofs for dwellings, published by the Timber Research and Development Association, which complies with *BS 5268-2:2002 Code of practice for permissible stress design, materials and workmanship*.

The joists are laid over the external supporting walls and connected to a wall plate by either skew nailing or by connector plates. The wall plate in turn is secured to the inner face of the wall by a 30 x 5mm galvanised mild steel or austenitic steel strap a minimum of 1m in length.

Self-assessment question 13.1

Why do the joists need to be secured to the supporting walls by a galvanised mild steel strap?

In addition, similar to the joists to upper floors, where the joists run parallel to an external wall, restraint straps should be placed at a maximum spacing of 2m and fixed over 3 adjoining joists, supported by *noggings*. The joist ends may finish flush with the external face of the wall and are covered with a *fascia board* (see Figure 13.5).

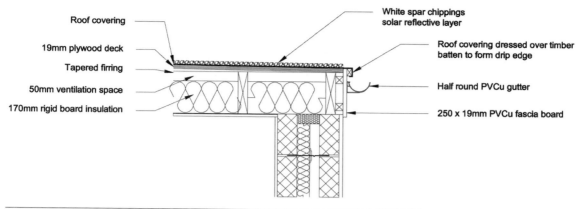

Figure 13.5 Eaves detail where the joists finish flush with the external wall

Self-assessment question 13.2

Why do the ends of the joists need to be covered with a fascia board?

Alternatively, joists may overlap the external face of the wall by a distance of 150–300mm to provide a *soffit* beneath, which is covered by a *soffit board* (see Figure 13.6).

Self-assessment question 13.3

Why might the joists be designed to overlap the external wall by a distance of 150–300mm?

The roof is not completely flat, but incorporates a slight gradient for the run-off of rainwater. It is recommended that the gradient should be a minimum of 1 in 40. Falls may be achieved by laying joists to the required slope, but this gives a sloping ceiling, which is considered to be unacceptable. More generally, falls are achieved by fixing tapered strips of timber, called *firrings*, to the tops of the joists (see Figure 13.5). These firrings are normally a minimum of 38mm thick for joists at 400 or 450mm centres and 50mm thick for joists spaced at 600mm centres.

The rainwater is generally drained to the edge of the roof, called the eaves, and collected by a gutter, which is fixed to the fascia board (see Figure 13.6). Alternatively, on longer span roofs, the rainwater may be drained to various collection points on the roof and collected within a hopper connected to a rainwater pipe running inside the building. It is very rare for flat roofs used in domestic construction to be large enough to require this latter facility. At the edges of the roof where the rainwater is not collected (the verges), a kerb is constructed to prevent

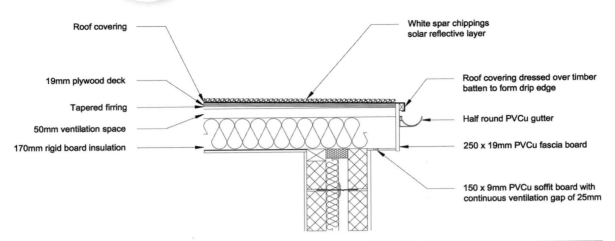

Figure 13.6 Eaves detail where the joists overhang the external wall

the rainwater pouring over the edge of the roof at these points (see Figure 13.7).

Where the roof adjoins an external wall to another building, such as where the roof is to an extension, an *abutment* is formed. A cavity tray needs to be incorporated within the abutment wall to prevent water seeping down behind the abutment and into the room below (see Figure 13.8).

The roof decking is fixed to the top of the joists and firrings. Suitable decking materials are similar to those for upper floors. If softwood timber boards are used it is advisable that the boards are a maximum of 100mm in width. If particleboard sheets are to be used they should be the moisture resistant category, preferably a cement bonded particleboard. Plywood sheets of use classes 2 or 3 provide a good decking for flat roofs as do oriented strand board of flaxboard sheets, provided the correct resin has been used in their manufacture. Strutting should be applied between joists where spans exceed 2.5m.

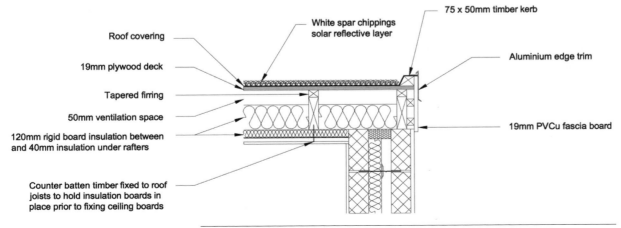

Figure 13.7 Verge detail

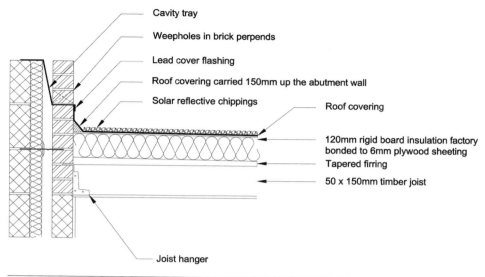

Cavity tray

Weepholes in brick perpends

Lead cover flashing

Roof covering carried 150mm up the abutment wall

Solar reflective chippings

Roof covering

120mm rigid board insulation factory bonded to 6mm plywood sheeting

Tapered firring

50 x 150mm timber joist

Joist hanger

Figure 13.8 Abutment detail

Thermal insulation may be provided by means of sheep's wool, cellulose fibre, glass fibre, mineral fibre or cellular plastics, such as expanded polystyrene fixed between the joists. Alternatively, a rigid insulation material such as cork, cellular plastics or low density fibreboard insulation board may be fixed over the decking. In some instances waterproof insulation materials may be fixed over the covering and weighted down to avoid being blown off the roof.

Self-assessment question 13.4

Why do insulation materials fixed above the decking need to be rigid?

Solid suspended concrete slab construction

A flat suspended concrete slab, spanning between supporting walls at roof level, may be cast as a flat roof structure. The slab comprises structurally designed mix concrete, incorporating reinforcement to prevent cracking in the lower tensile section, due to excessive deflection under load. The thickness of the slab will be designed according to the loads imposed and the span required; however a rule of thumb guide would be that the thickness should be a minimum of 1/30th of the span. The reinforcement may consist of a variety of rods or, alternatively, a mat of fabric mesh containing small diameter rods spot welded together in a square or rectangular matrix.

Self-assessment question 13.5

A concrete flat roof spans a distance of 4.500m. Using the rule of thumb method, what would be the minimum thickness of the roof slab?

When the concrete is first placed it is wet and has no inherent strength to support itself. It is therefore necessary to erect a temporary support framework, called *formwork* or *shuttering*, to provide the necessary support to the concrete whilst it is setting and achieving sufficient inherent strength to support itself between the supporting walls. The formwork generally consists of sheets of plywood to provide a deck onto which the concrete can be placed, supported by timber *bearers*, which in turn are supported by adjustable props (see Figure 13.9).

The formwork decking is levelled to coincide with the top edge of the supporting walls, and formwork is also placed around the edges of the proposed slab. If cavity walls are to be used for the supporting walls, the cavity will need to be closed at the top of the walls to prevent wet concrete from being poured down the cavity.

The reinforcement is laid on small concrete or plastic spacers to provide a 25mm gap between the formwork and the reinforcement so that the wet concrete can be placed underneath the reinforcement.

Self-assessment question 13.6

Why should it be necessary to raise the reinforcement from the formwork decking to enable the wet concrete to be placed beneath it?

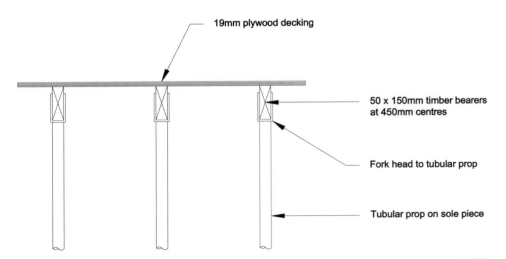

19mm plywood decking

50 x 150mm timber bearers at 450mm centres

Fork head to tubular prop

Tubular prop on sole piece

Figure 13.9 Formwork support to a concrete roof slab

Alternatively a concrete roof slab can be constructed using prestressed precast concrete beams supporting concrete blocks, similar to those used in suspended precast concrete ground floor construction. The beams are built into the supporting walls and have an inverted T-shaped profile. This enables the concrete blocks that provide the decking to be placed between the supporting beams.

Whether the concrete deck is formed from an in situ slab or precast concrete blocks and beams, the finished surface of the roof slab is flat. The required falls are provided by adding a finishing screed of a minimum thickness of 40mm and laid to the required gradient. The screed may incorporate lightweight aggregates in order to enhance the thermal insulation capabilities of the roof.

Combating interstitial condensation

Moisture is created by a number of activities taking place within a house. Cooking, washing clothes, bathing, heating, even breathing creates water vapour, which can be held in suspension in the air. *BS 5250:2011 Code of practice for control of condensation in buildings* estimates that for a house with five occupants, 7–14 litres of water vapour is produced each day.

Self-assessment question 13.7

What process turns water into water vapour that can be held in suspension within the air?

On top of this, new houses will have up to 4,000 litres of water incorporated into their construction, primarily as mixing water for concrete, mortar or plaster; in addition there will be moisture contained within the new timber, or rainwater may be absorbed by the materials within the fabric of the building due to exposure to the weather before the house became watertight. This water will evaporate into the atmosphere, as water vapour, over a period of up to a year following occupation of the house.

All air contains a certain amount of moisture, called *humidity*. The amount of moisture that air can sustain before it becomes saturated is related to its temperature. Generally the higher the temperature of the air the more moisture can be held in suspension. Conversely, the lower the temperature of the air the less moisture can be held in suspension and any excess must be precipitated out as water droplets, called condensation. The temperature at which the air becomes saturated with a quantity of water vapour is termed the *dew point temperature*.

A temperature gradient exists between the rooms of the house below the roof and the external environment above the roof. The materials of construction within the roof will affect the shape of this temperature gradient (see Figure 13.10).

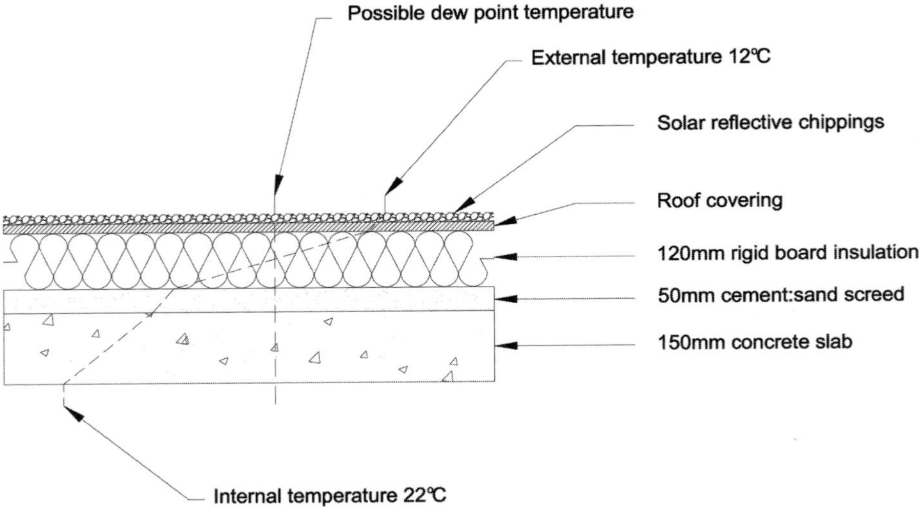

Figure 13.10 Typical temperature gradient through a flat roof

As air percolates through the roof construction it will cool and eventually reach its dew point temperature for the quantity of water vapour it is carrying. Any further drop in temperature below the dew point temperature will cause some of the water vapour to be precipitated out and condense into water droplets on cold surfaces.

Condensation may be of two types. The precipitated water vapour may condense on cold, non-absorbent surfaces such as ceramic tiles or window glazing. This is known as *surface condensation*. Alternatively the water vapour may condense as water droplets onto the materials of the structure of the building. This is known as *interstitial condensation*. Surface condensation is a nuisance and can cause mould growth on walls and ceilings if not treated, but it can be easily detected and remedied, generally by providing extra ventilation so that the heavily moisture laden air can be replaced by drier air to reduce its relative humidity. Interstitial condensation is more of a problem, because it occurs within the structure where it cannot be readily seen. The water droplets can be absorbed by the insulation materials, which will lower their thermal resistance values and make them poorer insulators or the water droplets may be absorbed by organic materials in the roof construction, which may cause them to decompose. Timber members in the roof will also become more susceptible to fungal decay.

There is therefore a need to prevent high moisture levels within the air from reaching a position within the roof construction where the dew point temperature may exist and interstitial condensation may occur.

There are three forms of flat roof construction that may be used to combat interstitial condensation:

- cold deck design
- warm deck design
- inverted roof design.

Cold deck design

A space of at least 50mm is left between the cold side of the thermal insulation layer and the structural deck, hence the name 'cold deck'. Air is allowed to pass through this space between the joists by providing a 25mm continuous air gap for ventilation at the roof edges (see Figure 13.6). Thus the air with a high water vapour content that is percolating through the roof space is replaced by drier air from outside. It is important to ensure that the insulation within the roof is carefully placed to avoid thermal bridging at the edges of the roof and still be able to maintain a gap for the air to flow above the insulation layer.

Self-assessment question 13.8

Why is the air gap above the thermal insulation layer and not below it?

A *vapour check* of polyethylene or aluminium foil at ceiling level helps to reduce the amount of water vapour contained within the air from entering the roof space. The materials used for vapour checks are impermeable to the passage of moisture vapour, but allow air to percolate through them. Although this system is quite advantageous it cannot be considered to be 100 per cent effective as it cannot be continuous across the roof.

Self-assessment question 13.9

Why cannot the vapour check be considered to be continuous across the roof?

The advantages of cold deck design are that the waterproof covering can be laid directly on top of the deck and is accessible for maintenance, repair and replacement when necessary. The ventilation will help to remove any water vapour that may have passed through the vapour check at ceiling level (see Figure 13.11).

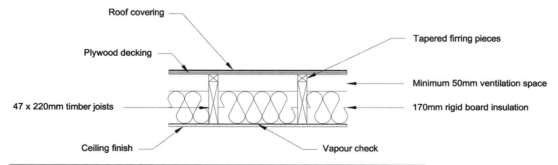

Figure 13.11 Cold deck design

The disadvantages of cold deck design are that it relies on natural ventilation to remove any water vapour that may have passed through the vapour check. Sometimes the weather conditions prevent natural ventilation from being able to ventilate the roof void adequately. Moreover, if the ventilation spaces are increased to improve ventilation efficiency, there is a likelihood that birds, insects or vermin could enter the roof space. In addition, if a large amount of insulation is required in the roof, the size of the joists will need to be increased to accommodate the extra depth of insulation and to still maintain an air void between the top of the insulation layer and the underside of the structural deck. This form of design can only be used with hollow suspended timber construction. In addition, the thickness of insulation required, coupled with the need to maintain a clear 50mm gap between the insulation and the underside of the decking for ventilation flow, tends to increase the depth of the joists.

Apart from the cost of the larger sized joists, this can also make the depth of the roof so large that it is not aesthetically pleasing. This problem may be overcome by using 120mm of cellular plastic insulation between the joists with 40mm of the insulation drawn under the joists. There is then the problem of securing the insulation to the underside of the joists before the plasterboard is fitted. This can be overcome by counter battening the roof to hold the insulation in place prior to the installation of the ceiling boards (see Figure 13.7).

Self-assessment question 13.10

Why can this form of design only be used with hollow suspended timber construction?

Warm deck design

The thermal insulation layer is placed on top of the structural deck. The deck is therefore on the warm side of the insulation, hence the name 'warm deck'. To prevent water vapour from passing through the thermal insulation layer, a *vapour control layer* is placed over the deck immediately beneath the insulation. The materials used for the vapour barrier can be the same as those used for the vapour check in cold deck design. The difference between the vapour control layer and the vapour check is that the former must be 100 per cent intact and sealed at all joints. Therefore it cannot be fixed mechanically to the deck, but will either be loosely laid on the deck or fixed by means of an adhesive. The waterproof covering is then laid on top of the thermal insulation layer (see Figure 13.12).

Self-assessment question 13.11

Why does the vapour control layer need to be placed beneath the thermal insulation layer and not above it?

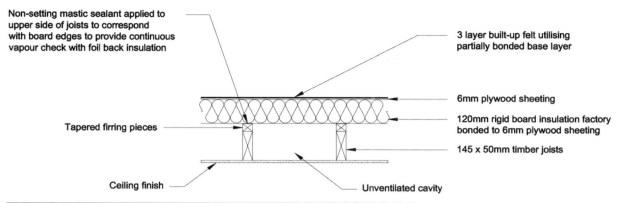

Non-setting mastic sealant applied to upper side of joists to correspond with board edges to provide continuous vapour check with foil back insulation

3 layer built-up felt utilising partially bonded base layer

6mm plywood sheeting

120mm rigid board insulation factory bonded to 6mm plywood sheeting

Tapered firring pieces

145 x 50mm timber joists

Ceiling finish

Unventilated cavity

Figure 13.12 Warm deck design

The advantages of warm deck design are that a void is not required in the roof construction, unless one is to be used for the passage of services. This form of design can therefore be used with either solid concrete slab or hollow suspended timber construction. Furthermore, ventilation is not required within the roof construction and the structural deck is kept warm by the insulation so that the materials used in its construction do not suffer from large fluctuations in temperature. The vapour control layer is able to be 100 per cent effective in preventing water vapour from entering the thermal insulation. A further advantage is that because the insulation is placed above the deck rather than within the roof void, smaller joist sizes can be used than in cold deck design.

The disadvantages of warm deck design are that the thermal insulation layer must be strong enough to support foot traffic for maintenance purposes. If the roof covering needs to be repaired or replaced the thermal insulation layer beneath may become damaged and also need to be replaced. In addition any failures in the waterproof covering may be difficult to detect.

Inverted roof design

Because the roof covering is impermeable it can also act as a vapour control layer. As the vapour control layer must be placed on the cold side of the thermal insulation, the insulation layer must be placed on top of the roof covering material. To prevent the lightweight insulation material from being blown off the roof, it is weighted down with ballast or paving materials (see Figure 13.13).

The advantages of inverted roof design are that there is no need for a separate vapour control layer. The waterproof covering fulfils the dual function of covering and vapour control layer. The thermal insulation layer also protects the covering from the vagaries of temperature variations and the possibility of being damaged by foot traffic.

The disadvantages of inverted roof design are that the thermal insulation layer is exposed to the external environment. It must be waterproof, as any absorption of water will affect its thermal performance.

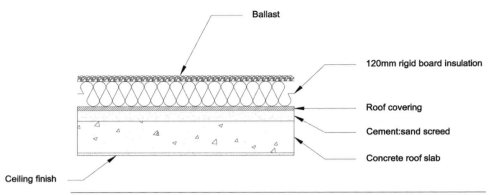

Figure 13.13 Inverted roof design

It must be frost resistant and resistant to degradation by ultra-violet light from the sun. It must also be weighted down to prevent it from being blown away by the wind. This increases the dead load on the roof and may require a stronger roof to be designed to accommodate this extra loading. The insulation layer must also be strong enough to withstand foot traffic on the roof. The roof design will need to allow rainwater to drain through gaps in the insulation layer so that it can be drained form the roof adequately. These gaps will need to be maintained regularly to ensure they have not become blocked with debris that may prevent adequate drainage from the roof. The gaps will allow some heat to escape from the roof. The thermal insulation layer may therefore need to be thicker to allow for this heat loss. The insulation layer may mask the waterproof covering beneath so that it becomes difficult to determine if deterioration of the covering has taken place. Any repair or replacement to the covering will require the thermal insulation layer to be removed first.

Although the inverted roof design can be used with both timber and concrete roof construction, it is more likely to be used on solid concrete slab construction due to the extra dead loading entailed.

Where concrete screeds are used and they are not given sufficient time to dry out before the impermeable covering is applied, moisture may become entrapped within the screed. This may be alleviated by the provision of vents embedded in the screed and protruding through the covering.

Flat roof coverings

Suitable coverings for flat roofs may be:

- mastic asphalt
- bitumen sheet built-up roofing or single ply sheeting using synthetic materials
- sheet metal coverings.

The coverings must provide the weatherproofing requirements for the roof and must be impervious, with sealed joints. The coverings will also need to be durable, as regular replacement is difficult.

The most popular flat roof covering for domestic construction has undoubtedly been bitumen sheet built-up roofing, mainly due to its relatively low cost in comparison to the other types of covering, but also because of its ease of application. However, modern synthetic materials, with high performance characteristics, are now being developed as single ply alternatives to bitumen sheet coverings. Additionally, the other two types of covering are occasionally used for domestic construction and they are therefore also considered here.

Mastic asphalt

Mastic asphalt has been considered before as a material for a damp proof course. It is a jointless, weatherproof and impermeable material. It is pliable and never completely rigid. It consists of a mixture of asphaltic cement, bitumen and inert mineral aggregates. It is heated on site in boilers to a temperature of 200^{0}C to become fluid; it is then hoisted to the roof and laid by wood float to cool and set in position.

It is laid in 2 coats to a minimum thickness 20mm horizontally and 3 coats to a minimum thickness of 20mm vertically to *BS 8218:1998 Code of practice for mastic asphalt roofing*. At the eaves it is dressed over the fascia board into an aluminium *edge trim* above the gutter (see Figures 13.5 and 13.6). At the verges it is dressed over a triangular timber kerb and terminated in an aluminium edge trim (see Figure 13.7). Where a timber kerb is used, expanded metal lathing is attached to the kerb to act as a key to the asphalt. On concrete roofs the verge may consist of an upstand kerb. To prevent the mastic asphalt cracking at the base of the upstand kerb an angle fillet is formed. At vertical abutments the asphalt is formed into an angle fillet and carried up the wall a minimum height of 150mm and dressed into a horizontal brickwork joint. A metal cover flashing is used to protect the covering (see Figure 13.8).

Although mastic asphalt is considered to be a jointless material it does have to be laid in bays, otherwise it will become too cold to be able to level effectively. The edges of each bay must therefore be warmed with the asphalt on the adjoining bay as it is laid, to ensure that no joint between bays occurs in the finished work. It is good practice to ensure that bay joints do not coincide in subsequent coats.

Horizontal asphalt is laid on an *isolating membrane* of impregnated flax felt to separate the asphalt covering from the roof structure. This prevents differential cracking from occurring, since the asphalt and the materials of the roof decking may well have different coefficients of moisture and thermal movement.

Mastic asphalt is more durable as a roof covering than bituminous felts, provided that it has been laid correctly. However, it is either derived from the distillation of crude oil or from natural deposits. Derivation from crude oil, pollution from its extraction and distillation causes harm to the environment and where it is extracted from natural deposits, damage can be done to the landscape and natural habitat. However, mastic asphalt is reclaimable and may be stripped from existing roofs, re-heated with a proportion of new material and laid again.

The laying of mastic asphalt is a specialist technique. Care has to be taken to ensure the asphalt is separated from the structural deck and that right-angled corners are avoided at verge and abutment details.

Bitumen sheet built-up roofing systems and single ply sheeting

Bitumen sheet roofing comprises a mat of glass fibres or synthetic polyester fibres rendered impervious to water by impregnation with an oxidized bitumen coating. *BS 8747:2007 Reinforced bitumen membranes for roofing. Guide to selection and specification* considers two types of bitumen sheet, Class 3 bitumen sheets with a glass fibre base, laid in 3 layers and Class 5 bitumen sheets with a polyester base laid in either 2 or 3 layers. The felts are supplied either with a sanded surface or a mineral surface. They are applied to the roof surface with hot bonding bitumen.

The first layer is laid at right angles to the direction of the fall and is partially bonded to the roof deck to allow for differential movement between the deck and the covering. Traditionally this was achieved by nailing the sheet to the deck at specific locations. This method has now been generally superseded by the provision of a grid of holes 10mm in diameter being cut in the sheet at the time of manufacture. When the hot bitumen sealing compound is poured and mopped over the sheet some of the bitumen compound will seep beneath the holes and bond to the deck beneath, whilst the rest of the base sheet remains detached from the deck. This overcomes the problem of having to nail the first layer of roof sheeting to the deck.

Self-assessment question 13.12

Why is it better to attach the first layer of roof sheeting by allowing hot bitumen sealing compound to seep beneath the holes in the sheet rather than nail the sheet to the deck?

Subsequent layers of roof sheeting are bonded to the first layer with hot bitumen sealing compound poured and mopped over the surface of the lower layer. Laps between adjacent sheets should be a minimum of 50mm in width.

At the eaves the felt is dressed over the fascia board into an aluminium edge trim above the gutter (see Figures 13.5 and 13.6). At the verges the felt is dressed over a triangular fillet and terminated in an aluminium edge trim above the kerb (see Figure 13.7). Where the felt abuts vertical walls it is dressed over an angle fillet and carried up the wall a minimum of 150mm and turned into a horizontal brickwork joint. It is protected by a cover flashing (see Figure 13.8).

Sheet coverings to flat roofs need to be able to withstand movement, particularly thermal movement during heating and cooling cycles, without breaking down or tearing. They also need to resist weathering.

Polyester-based sheets are generally up to ten times stronger and more flexible than glass fibre-based sheets and will therefore tend to be more durable and require less frequent replacement. They are however more expensive. It is estimated that a glass fibre-based sheet system will have a life span of between 7–15 years, whereas the polyester-based sheet system will have a life span of between 15–25 years.

Modern technological innovations have modified the bitumen binder to provide enhanced fatigue resistance up to 20 times that of the polyester-based sheets. These modified bitumen roofing sheets are available as a 3-layer styrene butadiene styrene (SBS) sheet system or a 2-layer atactic polypropylene (APP) sheet system. These allow the sheet membranes to be torched on to the roof. The sheets have an underside that is pre-treated with a covering of modified bitumen. When this covering is heated with a blow torch it softens and acts as an adhesive, allowing the sheet membrane to be stuck down to the roof surface. Some sheet systems can also be bonded with a polyurethane adhesive. It is estimated that the life cycle of these materials is between 20–30 years.

Bitumen products are extracted from crude oil. They are non-recyclable and have poor biodegradability.

Single ply sheet roofing systems have also been developed using synthetic products, such as Polyvinylchloride (PVC), Thermoplastic polyolefin (TPO) and Ethylene propylene diene terpolymer (EPDM). These products provide excellent resistance to weathering and thermal cyclic fatigue. They can be jointed using a hot air gun and damaged areas can easily be repaired. Again the life cycle of these materials is estimated at between 20–30 years.

Apart from their high durability, single ply membranes use less material than built-up roofing felts, however, like them, they are not easily recycled and derived from petrochemical resources, which have poor sustainability credentials.

Manufacturers of these modified bitumen and single ply high performance sheet membranes now require them to be installed on the roof by specialist contractors, as it has now become more likely that failure of the roof system will occur due to bad workmanship than poor materials.

Sheet metal coverings

These are produced from malleable metals in sheet form. The metals must be malleable (i.e. easily bent to shape without cracking) because the sheets will be required to adapt to various profiles when fitted on the roof. They have high durability and impermeability, better even than mastic asphalt. The materials are, however, expensive and they require plumbers to fit them. Typical metals used are lead, copper, aluminium and zinc. By far the most popular of these are lead and copper.

The metal sheets are laid over an isolating membrane of impregnated flax felt, similar to that used for mastic asphalt coverings, to overcome the problem of differential movement between the covering and the structural deck beneath. Sheet metals will have greater thermal movement than other flat roof coverings and so the joints between sheets need to be able to accommodate this movement.

Lead sheets for roofs come in 2.4m wide rolls of various lengths. Lead sheet is available in various thicknesses according to use category.

Lead sheet for flat roof coverings is normally of code nos. 5, 6, 7 and 8. Joints running parallel to the slope of the roof are formed by rolls and joints running at right angles to the slope are formed by drips (see Figures 13.14 and 13.15).

At the eaves the lead sheet is taken over the fascia board into the gutter. At the verges the lead sheet is dressed over an angle fillet and fixed to the top of the kerb. At abutments to vertical walls the lead sheet is dressed up the wall to a height of 150mm and turned into a horizontal brickwork joint and protected by a cover flashing.

Copper sheets for roofs come in 1800 x 600mm rolls of 0.45–0.60mm thickness. Joints running parallel to the slope of the roof are formed by rolls, and joints running at right angles to the slope are formed by welts (see Figures 13.16 and 13.17).

At the eaves the copper sheet is taken over the fascia board into the gutter. At the verges the copper sheet is dressed over an angle fillet and

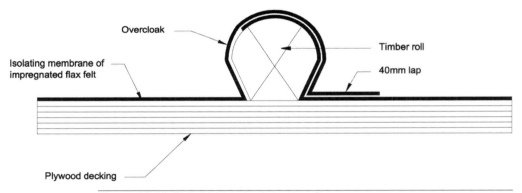

Figure 13.14 Rolled joint to lead sheet roofing

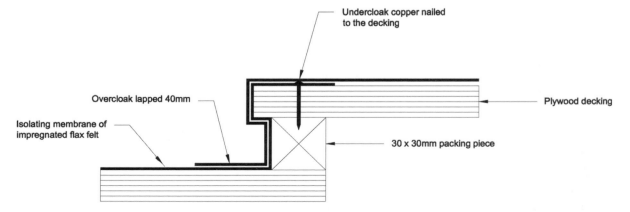

Figure 13.15 Drip joint to lead sheet roofing

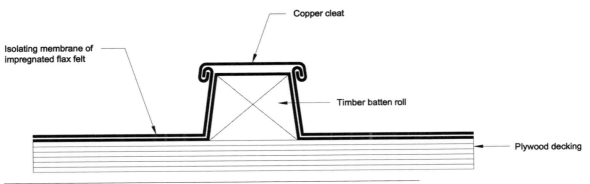

Figure 13.16 Rolled joint to copper sheet roofing

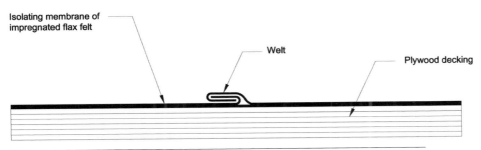

Figure 13.17 Welted joint to copper sheet roofing

fixed to the top of the kerb. At abutments to vertical walls the copper sheet is dressed up the wall to a height of 150mm and turned into a horizontal brickwork joint and protected by a cover flashing.

Both lead and copper sheet roof coverings are affected by oxidation and both metals form a thin protective film over their surfaces after being exposed to the air for several months. The lead oxide film produces a whitish-grey coating on the surface of the metal covering. The copper roof develops a green patina. Neither of these films is harmful to the metal covering and do help to protect the metal from further deterioration.

Metals are of course extremely durable, recyclable and reclaimable. However the extraction of the ore causes damage to the environment and the processing of the material has high embodied energy and also creates a large number of pollutants. Added to this, lead is a toxic material and must therefore be carefully disposed of. Because of the high value of metals, roofs using these materials are vulnerable to being stripped by thieves, leaving the building susceptible to water penetration damage.

Solar radiation

Flat roofs are exposed to the full force of the sun. The layer of thermal insulation incorporated into the roof construction prevents solar heat gain from being dissipated into the interior of the building. As a consequence

on hot, sunny days the roof covering may reach temperatures far in excess of normal working temperatures for the covering materials being used. This may cause a softening of the covering material and a breakdown in its structure. Mastic asphalt will begin to soften as it becomes hotter, this may cause vertical upstands to slump. Bitumen sheet coverings will suffer from the bitumen bonding compound beginning to liquefy and run and even the bitumen binder in the sheet itself to leach out. Also continuous heating and cooling of the roof sheet, causing the sheet to expand and contract during thermal cycles, can lead to fatigue within the sheet material, causing mechanical breakdown or tearing.

Heating up of the roof may also cause entrapped moisture to re-vapourise and attempt to pass through the roof covering to the external air, where the vapour pressure is much less. Because the coverings on flat roofs are impermeable, the water vapour will be unable to pass through easily and the vapour pressure differential between the entrapped water vapour and the ambient water vapour in the external air can cause blistering to occur in bituminous felt and mastic asphalt coverings. These blisters are a source of potential weakness in the covering and may be split when trodden upon by workmen engaged in roof maintenance.

In addition, ultra-violet rays present in sunlight may embrittle bitumen sheet coverings causing cracking to occur after later movement.

Suitable treatments may be white or aluminium paints applied to the surface of the covering to reflect the sun's rays off the roof surface. However, these will eventually wear off and need to be re-applied at regular intervals. Alternatively, white stone chippings may be embedded in the surface of mastic asphalt whilst it is still warm following application, or bonded to the surface of bituminous felt coverings with an application of hot bitumen compound. The use of white stone chippings on the roof surface also helps to improve the fire resistance of the roof against the ingress of fire from neighbouring properties. The chippings should not have sharp edges or they will puncture the covering when they are walked upon.

Bitumen sheets are available with a mineral surface finish but these are not really suitable for protecting the horizontal surfaces of flat roofs. They can, however, be used on upstands to kerbs and verges and abutments to vertical walls.

Self-assessment question 13.13

Why do the white stone chippings need to be embedded or bonded to the surface of the covering?

Single ply roof sheeting can be supplied in a variety of colours and light colours will act as an effective solar reflective surface. EPDM sheets have been proven to have good resistance to ultra-violet radiation.

Visit the companion website to test your understanding of Chapter 13 with a multiple choice questionnaire, and to see the following features illustrated with full-colour photos:

- **13.1** A welted joint to metal sheet roofing
- **13. 2** Bitumen felt flat roof with solar reflective chippings
- **13.3** Insulation to cold deck flat roof

CHAPTER **14**

Pitched roof construction and coverings

Introduction

Pitched roofs have their surfaces sloping at an angle greater than 10^0 to provide a run-off for rainwater at the eaves. The roofs may be pitched on one side only (*monopitch*), on two sides (*dual pitch*) or on three or four sides (*hipped*). In addition, pitched roofs may be designed where the lower portion of the roof is pitched at a steeper angle than the upper portion (*mansard*). On dual pitch roofs the two ends of a detached building that are not covered by the roof structure are termed the *gable ends*.

This chapter will consider the formation of pitched roofs for domestic buildings commencing with the simple *couple roof* and progressing through the *close couple roof*, the *collar roof*, the *double roof*, the *triple roof*, and the *trussed rafter roof*. Finally the *hipped roof*, the *valley roof* and the *lean to roof* are discussed.

Coverings to pitched roofs are normally in the form of small tile units, fixed to *battens*, which in turn are fixed to the rafters. Weatherproofing of the covering is provided by overlapping the tiles at their heads and bonding the tiles at their sides to provide a *double lap*. Alternatively, tiles may be *single lap* and use an overlapping joint at the side or a patent edge fixing to adjacent tiles.

This chapter considers the main materials used in pitched roof coverings for houses and their method of installation. It also considers special tiles that are manufactured for a variety of special locations or uses. Finally it considers the topic of flashings to pitched roofs, the materials used and the method of installation.

The couple roof

This is the simplest form of pitched roof construction and is suitable for short span roofs up to approximately 3m span. Timber *rafters* bear onto 100 x 75mm *wall plates* at their base and are fixed at 400, 450 or 600mm centres to a 175 x 32mm *ridge board* at their heads (see Figure 14.1).

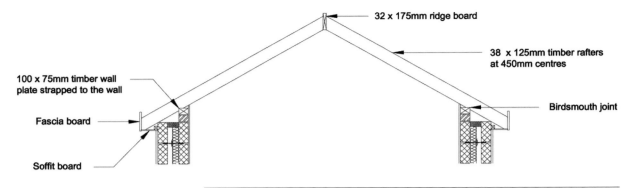

Figure 14.1 A couple roof

The wall plates are fixed to the top of the supporting walls with the aid of galvanised mild steel or austenitic steel straps at a maximum of 2m centres. Where wall plates are joined in their length a half-lapped joint is used. Where a cavity wall is used for the supporting wall the top of the cavity wall should be closed off and the wall plate positioned on top of this closing.

Self-assessment question 14.1

Why should the wall plate be positioned above the closing at the top of a cavity wall?

The foot of the rafter is cut with a *bird's mouth joint* to fit over the wall plate and nailed to the wall plate or fixed with framing anchors (see Figure 14.1). The rafters will generally extend beyond the wall to provide an overhang. A soffit board is fixed to their underside and a fascia board is fixed to their end (see Figure 14.1). The heads of the rafters are cut to provide a perpendicular joint with the ridge board and then *skew nailed* to each side of the ridge board.

The sizes of the rafters are determined by the span, their spacing and the superimposed load. Section sizes and spans for rafters and purlins are now provided in Eurocode 5 span tables: for solid timber members in floors, ceilings and roofs for dwellings, published by the Timber Research and Development Association, which complies with *BS 5268-2:2002 Code of practice for permissible stress design, materials and workmanship*.

Vertical strapping at eaves level is not required if the pitch of the roof is 15⁰ or more and the roof is tiled or slated. However, it may be required in severely exposed areas where wind gusts can be problematic. Tension straps located at not more than 2m centres are required to tie at least three end rafters to the gable wall. This strap should be turned over and built in to the inner leaf of the cavity wall (see Figure 14.2)

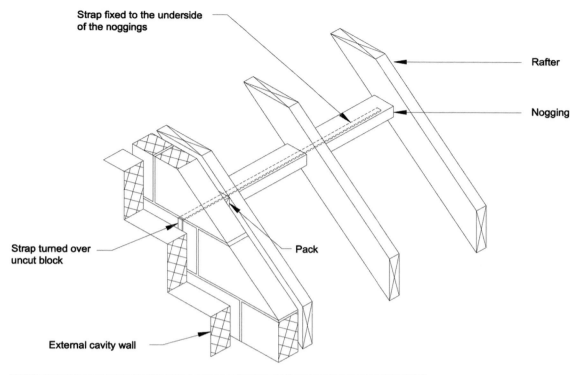

Strap fixed to the underside
of the noggings

Rafter

Nogging

Strap turned over
uncut block

Pack

External cavity wall

Figure 14.2 Strapping at a gable wall

The close couple roof

Where spans exceed 3m the dead load and superimposed loading on the rafters could cause them to 'spread' at their base, thus imparting an inclined thrust to the supporting walls. To prevent this occurring the paired rafters on each side of the roof may be joined together at their base by a horizontal timber called a *ceiling joist* (see Figure 14.3). Eurocode 5 span

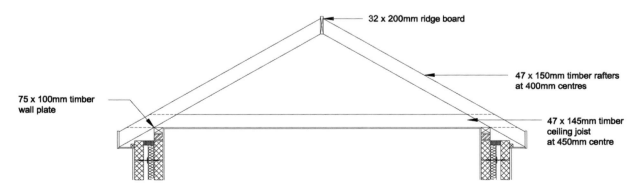

32 x 200mm ridge board

47 x 150mm timber rafters
at 400mm centres

75 x 100mm timber
wall plate

47 x 145mm timber
ceiling joist
at 450mm centre

Figure 14.3 A close couple roof

tables for solid timber members in floors, ceilings and roofs for dwellings, referred to earlier, provides suitable sizes for ceiling joists, dependent upon the dead load of the roof, the span of the joist and the joist spacing.

Self-assessment question 14.2

Why is the close couple roof more structurally stable than the couple roof?

The ceiling joists perform another useful function in that they can be used to support the ceiling finish on their underside. In some properties the floor to ceiling height in the upper storey may not be sufficient to allow the ceiling joist to be fixed at the base of the rafters. Where this occurs the ceiling joist may be raised up the roof to form a *collar* and to provide an increased ceiling height. However, the higher up the roof the collar is situated the greater will be the thrust imposed by the rafters on the supporting walls. The collar should therefore not be positioned any higher than one-third up the total height of the roof (see Figure 14.4).

At spans in excess of 3.5m the ceiling joists may show signs of excessive deflection. In order to prevent this *binders* are normally fixed at right angles to the top of the joists at mid-span (see Figure 14.5). Sizes of binders are given in Eurocode 5 span tables for solid timber members in floors, ceilings and roofs for dwellings and are dependent on the dead load of the roof, the span of the binder and the spacing of the ceiling joists.

At spans in excess of 4m the binders themselves may need support to prevent excessive deflection. This can be provided by 50 x 100mm *hangers* fixed vertically between the binders at their feet and the ridge board at their head at every fourth ceiling joist (see Figure 14.6).

Self-assessment question 14.3

What type of structural force is the hanger subjected to?

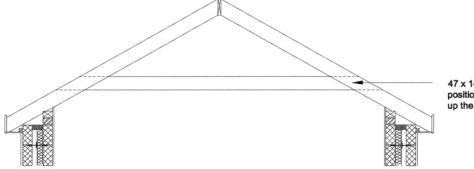

47 x 145mm collar at 450mm centres positioned a maximum of one-third up the height of the roof

Figure 14.4 A collar roof

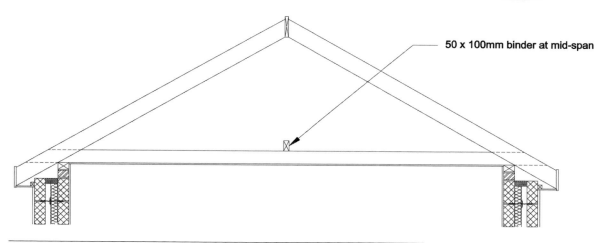

Figure 14.5 A binder strengthening the ceiling joists

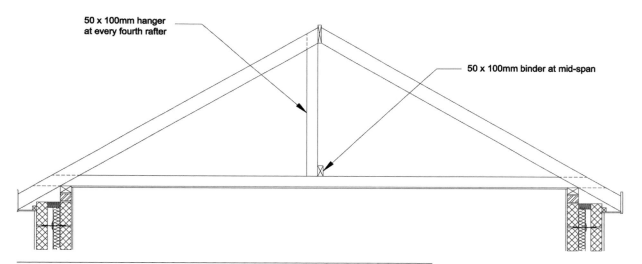

Figure 14.6 A hanger supporting the binder

The double roof

Where roof spans exceed 6m the rafters will require support in order to prevent them deflecting excessively. This support is provided by the *purlin*, positioned normal to the rafters at mid-span, creating a *double roof* (see Figure 14.7). The size of the purlin is considered in Eurocode 5 span tables for solid timber members in floors, ceilings and roofs for dwellings, referred to earlier.

The purlins span between *gable end* walls on dual pitched roofs, but are likely to require further support to prevent excessive deflection on long spans. This support may be provided by *inclined struts*, notched to the underside of the purlins and supported on internal loadbearing walls

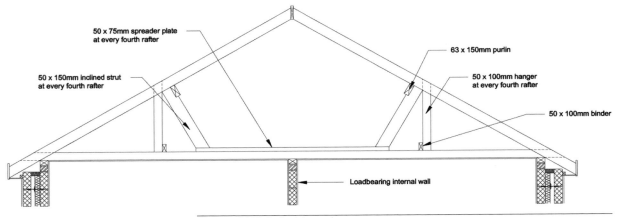

50 x 75mm spreader plate at every fourth rafter

63 x 150mm purlin

50 x 150mm inclined strut at every fourth rafter

50 x 100mm hanger at every fourth rafter

50 x 100mm binder

Loadbearing internal wall

Figure 14.7 A double roof

at every fourth pair of rafters (see Figure 14.7). Where rafters need to be joined in their length a *scarf joint* is used (see Figure 14.8) positioned centrally over the purlin. Where purlins need to be joined in their length a similar joint is used positioned above the inclined struts.

If loadbearing support can be positioned directly beneath the purlin then the purlin may be positioned vertically and cut into the rafters it supports (see Figure 14.9).

Where the vertical support to the inclined struts or purlins has to be provided by internal loadbearing walls, the internal room planning of the house may be restricted. Where this support to the purlins is not available or not desirable a specially constructed timber truss may be used to support the purlins. This is called a *triple roof*.

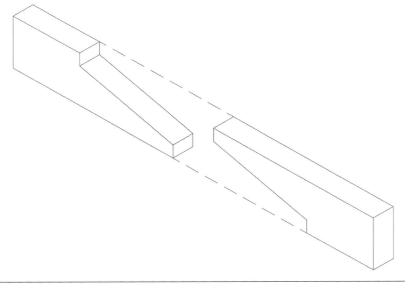

Figure 14.8 A scarf joint

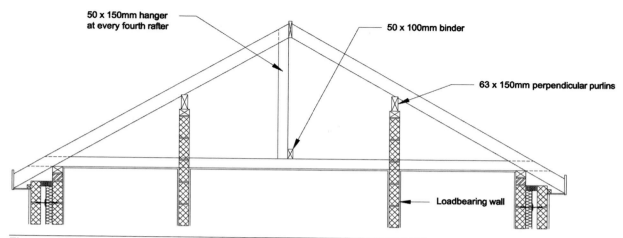

Figure 14.9 Vertical support to a purlin

The triple roof

This comprises a *truss* consisting of rafters, *struts*, *ties*, hangers and ceiling joist (see Figure 14.10) prefabricated into a self-supporting framework and spaced at 1.8m centres to support the purlins and having common rafters spaced between them. The whole roof is then tied together at the ridge, purlins and binders (see Figure 14.10).

The members of the truss are connected together by means of bolts through *toothed plate* or *split ring connectors*. The trusses may either be completely prefabricated at the works, transported to site and lifted into position on site or assembled on site from their pre-made components before being erected into position.

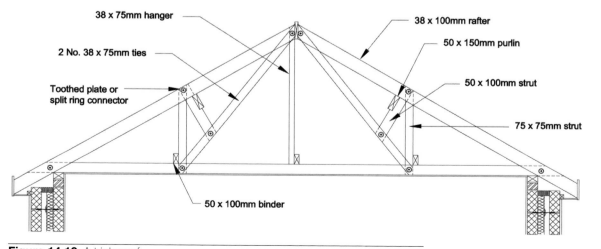

Figure 14.10 A triple roof

The trussed rafter roof

The principle of triangulation exemplified in the trusses used in triple roof construction has been further developed to create *trussed rafter* construction. Trussed rafters are completely prefabricated trusses manufactured from strength graded timber, butt jointed together with the aid of galvanised mild steel *nail plate connectors*. The truss comprises rafters, struts, ties and ceiling joist (often called a ceiling tie) in a 'fink' or 'fan' configuration (see Figure 14.11).

The trusses are completely self-supporting, spaced at 600mm centres, replacing the common rafters in a traditional 'cut' roof and dispensing with purlins, hangers, binders and ridge boards (see Figure 14.11). Because the trusses are prefabricated the spans and pitches must be standard and uniform.

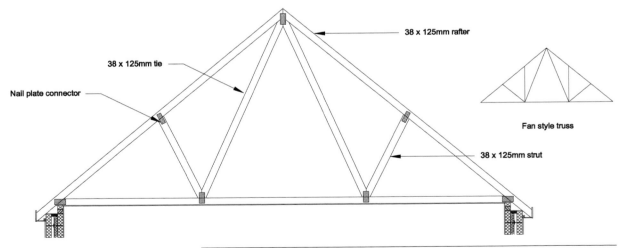

Figure 14.11 A trussed rafter

Self-assessment question 14.6

Apart from the drawback of having to accept standardised spans and pitches, what other disadvantages could there be to using prefabricated trusses in the roof?

There are other configurations that are available from trussed rafter manufacturers and one of the most popular is the room in the roof truss. Most of these trusses can be designed to span between the front and rear walls of a house (see Figure 14.12). However, if loadbearing walls are located near the centre of the span of the truss, greater room widths within the roof space will be possible.

There are two concerns with room in the roof trusses that do not affect normal fink or fan trusses. The first is the position of the thermal insulation. This should be placed above the ceiling joist then taken up the side of the vertical member in the truss which will form the wall of the room within the roof space, then taken over the top of the top ceiling tie which forms the ceiling to the room. An air gap of 50mm should be provided between the top of the insulation and the underside of the roof covering and this may affect the depth of the rafter required for this type of roof. In addition vents at eaves level, equivalent to a continuous 25mm air gap, will need to be provided (see Figure 14.13).

It is possible to place a 60mm rigid insulation board between the vertical timbers which form the cheek walls of the room. The non-

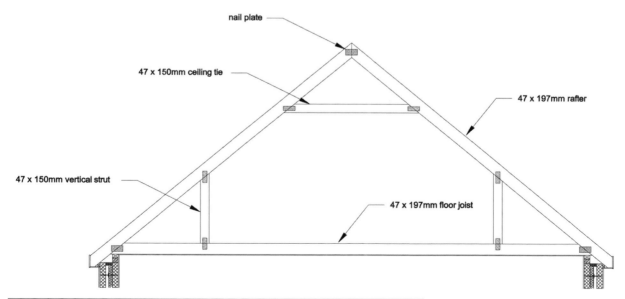

Figure 14.12 An attic trussed rafter

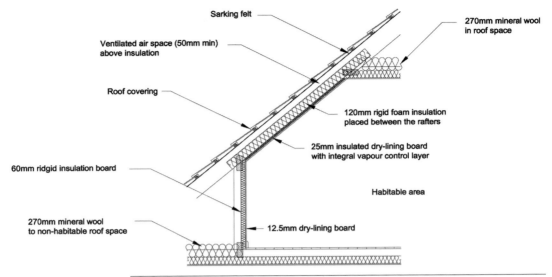

Sarking felt

270mm mineral wool in roof space

Ventilated air space (50mm min) above insulation

Roof covering

120mm rigid foam insulation placed between the rafters

25mm insulated dry-lining board with integral vapour control layer

60mm ridgid insulation board

Habitable area

270mm mineral wool to non-habitable roof space

12.5mm dry-lining board

Figure 14.13 Thermal insulation to a room in the roof

habitable area of the roof behind the walls needs to be insulated at ceiling level with 270mm mineral fibre wool. There are two main ways of insulating the pitched area of the roof. They consist of either rigid foam insulation between and under the rafters or alternatively between and over the rafters. The first method would be made up of 120mm insulation between the rafters with 12.5mm plasterboard bonded with 25mm insulation fixed to the underside of the rafters. There should be a minimum of 20mm air gap above the insulation to enable the breathable sarking membrane to drape between the rafters in an unventilated roof. If sarking felt is to be used in a ventilated roof system then a minimum 50mm air space needs to be maintained above the insulation.

The second option would be to use 60mm rigid insulation between the rafters and a second layer of 65mm insulation above the rafters. A breathable sarking membrane should then be placed directly above the insulation. This method of insulating then requires a 38 x 38mm counter batten to be fixed directly over the rafters to enable the fixing of the slate battens.

The between and under method is best suited to loft conversions, as it does not require the existing roof covering to be completely removed. However it does lower the ceiling height on what already may be a low ceiling. Another disadvantage of this system is that the existing rafters may not be of sufficient depth to accommodate the 120mm insulation, so the rafter may need to be counter battened to create the depth, keeping in mind the requirement for air space above the insulation.

The between and over method is best suited to new buildings, where it is easier to place the second layer over the rafters prior to fixing the roof covering.

The second concern relating to this type of truss is the need for fire resistance to the floor construction of the room in the roof truss. This is

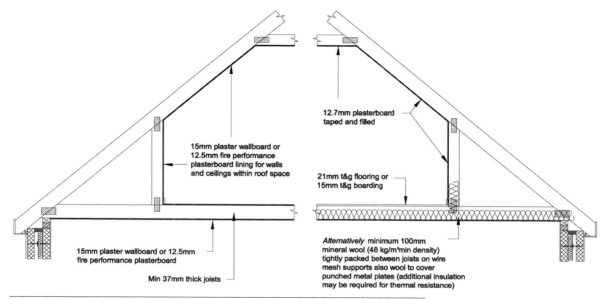

12.7mm plasterboard
taped and filled

15mm plaster wallboard or
12.5mm fire performance
plasterboard lining for walls
and ceilings within roof space

21mm t&g flooring or
15mm t&g boarding

15mm plaster wallboard or 12.5mm
fire performance plasterboard

Min 37mm thick joists

Alternatively minimum 100mm
mineral wool (48 kg/m³min density)
tightly packed between joists on wire
mesh supports also wool to cover
punched metal plates (additional insulation
may be required for thermal resistance)

Figure 14.14 Fire resistance methods to an attic trussed rafter

because the accommodation created within the roof void constitutes an extra storey to the house, which requires the floor construction (i.e. the base of the truss) to have a 30 minute fire resistance. It is recommended that the ceiling joist to these trusses is increased to 47mm thickness and to install a mineral wool insulation of 48kg/m³ minimum density and a minimum of 100mm thickness within the floor construction, taken at least 50mm up the vertical member each side of the room and covering the metal plate fasteners at the floor level. In addition the floor decking should be either 21mm thick tongued and grooved boarding or 15mm thick tongued and grooved chipboard (see Figure 14.14). Alternatively, a similar fire resistance to the floor can be achieved by the use of a higher fire resistance grade plasterboard fixed to the underside of the floor (the ceiling of the room below).

To avoid distortion and to prevent damage, trussed rafters should be carefully stacked on lorries for delivery to site and stacked clear of the ground and protected from the weather when on site. The trusses require stabilising during erection, using battens, and require internal diagonal wind bracing of a minimum size of 100 x 25mm on completion. *BS 5268-3: 1998 Code of practice for trussed rafter roofs* gives recommendations on the positioning of this wind bracing. This needs to be carefully considered where room in the roof trusses are used, otherwise the bracing may restrict the useable space created within the roof void. The trussed rafters are fixed to the wall plates using galvanised metal *truss plates* (see Figure 14.15).

Restraint straps are also needed to provide stability to the gable walls and hold down the roof against wind uplift. The straps should be fixed to at least three trusses and supported by timber *noggings* between the trusses, as considered previously (see Figure 14.2).

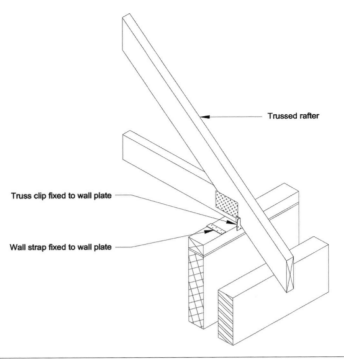

Figure 14.15 Truss plate fixing to a trussed rafter

Where the roof is designed to provide an overhang at the gable ends a *gable ladder* is constructed. This consists of horizontal noggings fixed between the last trussed rafter on the inside of the gable wall and extending through the gable wall to provide support to the *barge board* at the extremity of the overhang (see Figure 14.16). The gable ladder should be no wider than the spacing between the trussed rafters.

Where water tanks are located in the roof they should be mounted upon stands that spread the load to the *node points* (the points where truss members are connected) over at least three trusses.

Hipped roof

Hipped roofs are roofs that are pitched on three or four sides, comprising common rafters, hip rafters at the external intersection of pitched surfaces and short jack rafters, spanning between the wall plates and the hip rafters (see Figure 14.17).

Because hip rafters need to provide a fixing for the jack rafters, they need to be deeper than common rafters. To prevent the wall plates spreading at the corners of the roof immediately beneath the hip rafters, timber angle ties should be fixed across the corner and secured to the hip rafter by means of a galvanised mild steel dragon tie (see Figure 14.18). Purlins will need to be mitred at the hips.

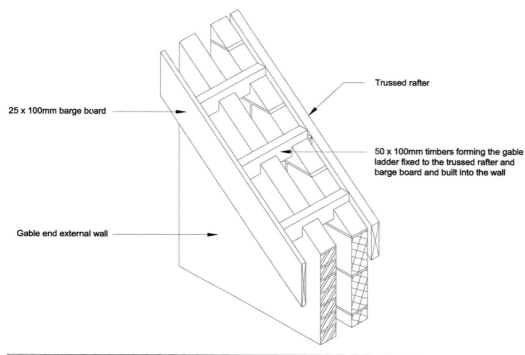

25 x 100mm barge board

Trussed rafter

50 x 100mm timbers forming the gable ladder fixed to the trussed rafter and barge board and built into the wall

Gable end external wall

Figure 14.16 A gable ladder and barge board

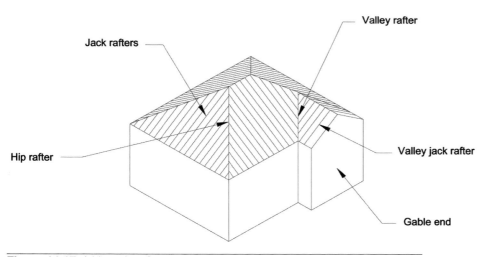

Jack rafters

Valley rafter

Hip rafter

Valley jack rafter

Gable end

Figure 14.17 A hipped roof

Valley roof

At all internal intersections of roofs, valleys are constructed. These are formed in a similar way to hips and have a valley rafter at the intersection of the pitched surfaces, with valley jack rafters to each side, spanning

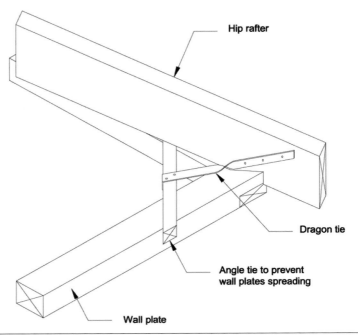

Figure 14.18 A dragon tie

between the valley rafter and the ridge board. Once again valley rafters need to be deeper than common rafters due to their need to support the end of the valley jack rafters.

Both hipped roofs and valley roofs can be constructed using adapted trussed rafters.

Lean to roof

These are roofs that are pitched on one face only (monopitch) and rely on an abutment wall for support. The rafters bear on a supporting wall at their base and an abutment wall at their head, via a wall plate fixed to the abutment wall.

Where longer spans are required, lean to roofs may be constructed from monopitch trussed rafters.

Ventilation

Similarly to flat roofs, pitched roofs can suffer from interstitial condensation. As with the cold deck design it is virtually impossible to provide a completely effective vapour control layer at ceiling level and so pitched roofs need to be adequately ventilated. This can be achieved by providing a continuous gap at eaves level on opposite sides of the roof space. This gap should be a minimum of 10mm wide for roofs with a pitch greater than 15°, and 25mm wide for roofs of 10°–15° pitch or where

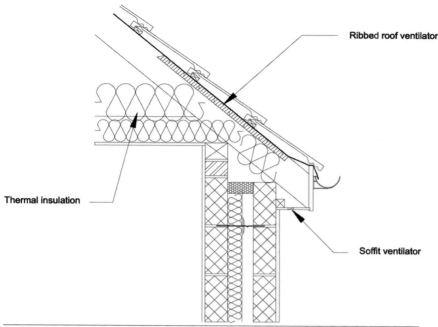

Figure 14.19 A ribbed roof ventilator

the thermal insulation layer is located above ceiling level. A mesh to prevent the entry of birds, insects or vermin should protect the gaps at eaves level. Alternatively, purpose made ventilation grilles can be fitted into the soffit board at regular intervals to equate to the same amount of ventilation as would be provided by a continuous gap. These grilles can incorporate suitable mesh.

Where the thermal insulation layer is brought to the eaves level, proprietary roof ventilators, consisting of a ribbed plastic board, can be placed over the thermal insulation, allowing the air from the ventilation gap to flow unimpeded into and out of the roof space (see Figure 14.19).

Pitched roof coverings

The pitch of the roof is determined by the style of the roof tile and the material of its manufacture. Roof tiles used in domestic construction are made from three main materials:

- slate
- clay
- concrete.

These are available in a number of different styles and sizes.

To prevent wind driven rain from entering the roof under the tiles, a secondary weatherproof barrier is provided using *sarking membrane*; this can either be a bituminous felt, similar to the bitumen sheet used for flat roof coverings, but is reinforced with hessian to prevent it cracking where it is unsupported between the rafters, or alternatively, proprietary products have been developed based on strong, reinforced plastic sheeting to provide the under-tile protection to the roof. The sarking membrane is draped over the rafters and lapped at intersections between 100mm–225mm dependent on the pitch. The tile battens are nailed on top and the sarking material is dressed into the rainwater gutter at the eaves.

On some roofs a triangular *sprocket* is fixed over the rafters at the eaves. This supports the underlay and also enables the last course of tiles to be raised so that they can drain into the rainwater gutter.

On some older roofs that were located in particularly exposed positions, roof boarding was placed over the rafters in order to overcome the problem of wind entering the roof beneath the coverings. Where this form of construction is used it is necessary to provide *counter battens* running up the slope of the roof at a similar spacing to that of the rafters, to enable the tiling battens to be raised above the surface of the sarking felt.

Self-assessment question 14.7

Why do the counter battens need to raise the tiling battens above the surface of the sarking felt in boarded roofs?

Slates

Slates are derived from metamorphic rocks having a laminar structure, which allows the rock to be split into tiles of 3–11mm in thickness, dependent on the type of slate. Welsh slates are blue-black in colour whilst Westmoreland slates range from olive green to pale green in colour. Because slates are derived from rocks they provide a very durable covering and can be reclaimed for re-use. They have low embodied energy, but their extraction can lead to landscape degradation. New slates are expensive, due to limited supplies in the UK. Less expensive slates are now being imported from sources in China, although these have high energy used in their transportation. A synthetic alternative is available and is much lighter in weight as well as being less expensive than traditional slate, but has embodied energy from its manufacture. Interlocking recycled slate tiles are now also being produced, which comprise 60–80 per cent recycled slate and can be used on low pitched roofs while maintaining the look of natural slate.

BS EN 12326-1:2004 Slate and stone products for discontinuous roofing and cladding. Product specification stipulates sizes for slates ranging from 254 x 152mm (Units) to 660 x 406mm (Empresses). Different sizes of slate are given different names, such as Countesses, Marchionesses and Duchesses. Small slates require steeper pitches than larger slates. Pitches may range from 22½°–45°. Slates are either head or centre nailed to the battens using copper nails. Every slate needs to be nailed to the tiling batten. The nail holes are punched with a special tool a minimum of 25mm from each edge with head nailing and halfway down the slate and 25mm from the edge for centre nailed slates. Centre nailing is preferred to head nailing as it causes less 'rattling' in high winds and reduces the chances of fixing nails being prised out of the battens.

Self-assessment question 14.8

Why should centre nailing be better than head nailing for reducing wind rattling of the slates and reducing the chances of fixing nails being prised out of the battens?

Slates are lapped at their heads by 65mm for a pitch of 40^0, 75mm for a pitch of 30^0, 90 mm for a pitch of 25^0 and 100mm for a pitch of 22½^0. The tiling battens are spaced at a distance termed the *gauge* (see Figure 14.20). The gauge is determined by:

$$\text{Gauge} = \frac{\text{length of slate} - (\text{head lap} + 25\text{mm})}{2}$$

Self-assessment question 14.9

What would be the gauge for a roof pitched at 30^0 and covered with Viscountesses (406 x 254mm)?

At the eaves an under eaves course of slate should be used (see Figure 14.21).

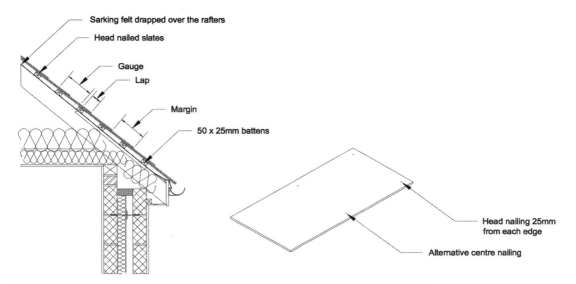

Sarking felt drapped over the rafters
Head nailed slates
Gauge
Lap
Margin
50 x 25mm battens
Head nailing 25mm from each edge
Alternative centre nailing

Figure 14.20 Slate covering to a pitched roof

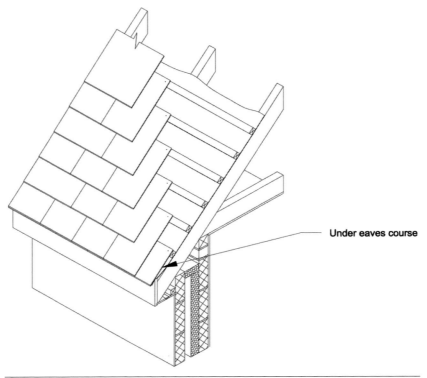

Under eaves course

Figure 14.21 Bonding of slates in successive courses

Because slates do not have a side lap to adjoining tiles they must be overlapped in successive courses (i.e. they are a double lap tile). The bond created will be equivalent to half the width of the slate. In order to create this bond the edge slates in alternate courses will need to be 1½ times wider than the other slates (see Figure 14.21). Some designers consider it to be aesthetically pleasing to use diminishing sizes of slate on a roof. This type of roof covering has larger sizes of slates at the bottom of the roof (the eaves) and the sizes of slate diminish up the roof, terminating in the smallest sizes of slates at the top (the ridge).

Although slates are very durable they can suffer from delamination if they are exposed for long periods to the effects of acid rain. This may also cause the 25mm border between the nail holes and the top of the slate in head nailed slates to be eroded, causing the slate to come free of its fixings and slide down the roof.

Clay tiles

Clay tiles are manufactured from either handmade or machine pressed clay, and kiln fired in a similar manner to bricks to *BS EN 1304:2005 Clay roofing tiles and fittings. Product definitions and specifications*. They may have pigments added to create different colours and may have a smooth

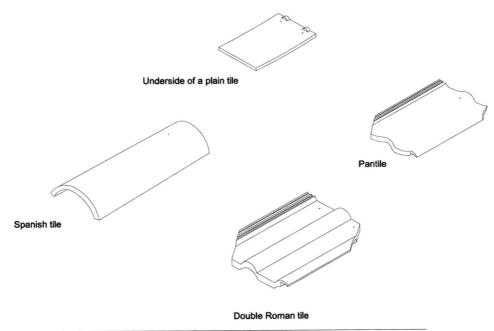

Figure 14.22 Styles of clay tiles

or a sand faced finish. They are manufactured in a number of different types (see Figure 14.22):

- Plain tiles (sometimes called Rosemarys)
- Pantiles
- Roman tiles
- Spanish tiles.

Plain tiles are the commonest form. They are available in 265 x 165mm size with thicknesses of 10–15mm. Because like slates they are double lapped tiles and therefore need to be butt jointed to adjoining tiles and bonded in courses, tile and a half tiles are also available.

The tiles incorporate nibs on their undersides, which enable them to be hooked over the tile batten so that they do not need to be fixed at every course. Nailing is therefore undertaken at every fourth course of tiles and also at areas of the roof where wind forces are particularly severe (i.e. at the eaves, the verge, and the ridge and around upstands such as chimneys).

Because clay tiles absorb water they must be laid to steeper pitches than other tiles, to allow water to run off the roof more quickly, and require three tile thicknesses to be provided at the head lap and two tile thicknesses between laps. For a 40⁰ pitch, the tiles would have a head lap of 65mm and a gauge of 100mm (see Figure 14.23). As with slates an under eaves course of tiles should be used. The tiles are cambered in their length to resist capillary attraction of water between tiles (see Figure 14.23).

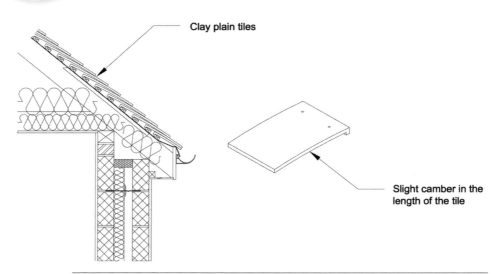

Clay plain tiles

Slight camber in the length of the tile

Figure 14.23 Eaves detail for clay tiles

Self-assessment question 14.10

How does cambering the tile resist capillary attraction of water between tiles?

Pantiles are 355 x 255mm in size. They have a waved profile and, unlike plain tiles, are single lapped and are laid with overlapping side joints, which are continuous from the eaves to the ridge. They therefore do not need to be bonded in courses. The head lap is normally 75mm and to prevent too great a build-up of thickness where the tiles overlap at the head and at the sides, two opposite corners of the tiles are chamfered (see Figure 14.22). They may have a smooth, sanded or glazed finish.

Roman tiles are 420 x 345mm in size and like pantiles are also single lapped with opposite corners chamfered. They comprise a flat portion with either one or two rolls to form the side lap (see Figure 14.22). They may have a smooth, sanded or glazed finish.

Pantiles and Roman tiles may also be manufactured with interlocking side joints. This consists of grooves on one side edge of the tile and a corresponding nib on the other side edge. Adjoining tiles interlock by the nib on one tile fitting into the groove of the adjoining tile (see Figure 14.22). Grooves on the underside of the lower edge of the tile prevent capillary attraction of rainwater between tiles at the head lap.

Spanish tiles are 355 x 230mm in size. They consist of convex upper and concave lower tiles (see Figure 14.22). The upper tiles taper down from the head to the tail and are fixed to battens running up the slope of the roof, rather than across the roof as battens for other tiles do. The lower tiles are skew nailed through their edges to the battens. Spanish tiles, as the name suggests, are commonly used as a roof covering in Spain and other Mediterranean countries, but are rarely used in the United Kingdom.

Clay tiles can suffer from frost damage if rainwater is not allowed to run off the roof quickly and is then absorbed by the tiles and freezes in cold weather. Frost damage will cause delamination of the tile. It is therefore important to ensure that roofs covered with clay tiles are laid to a steeper pitch than that required for slate or concrete tile coverings.

Clay tiles have similar sustainability credentials to clay bricks. The extraction of the clay causes landscape degradation and they have relatively high embodied energy through their method of manufacture. However, they are a home produced product, so transportation energy costs are lower than alternative roofing materials that have to be imported. They are also durable and can be reclaimed from demolition for re-use.

Concrete tiles

Concrete tiles are made from a fine mixture of cement and well graded sand to *BS EN 490:2011 Concrete roofing tiles and fittings for roof covering and wall cladding. Product specifications*. They are machine made by placing the mixture into steel moulds and curing for 1–2 days. Colours can be achieved with the addition of a pigment to the mix. Most concrete tiles have a sand faced finish although smooth finishes are also available. They are available as plain tiles, pantiles and Roman tiles in similar sizes to their clay equivalents. They are also available as interlocking tiles (see Figure 14.24).

Concrete tiles tend to be more durable than clay, since no firing is used in their manufacture and the tiles are not of a laminar structure.

The sustainability credentials of concrete tiles are similar to those for concrete bricks and blocks. They have a high amount of embodied energy due to their method of manufacture and the quarrying of constituent materials leads to landscape degradation. However, they have high durability and relatively low transportation energy costs. They are also now being produced with recycled aggregate content.

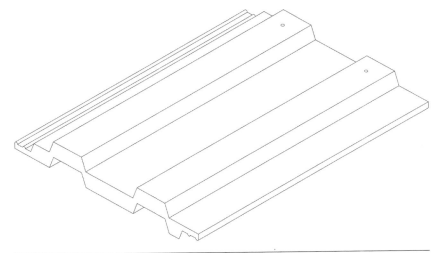

Figure 14.24 Interlocking concrete tiles

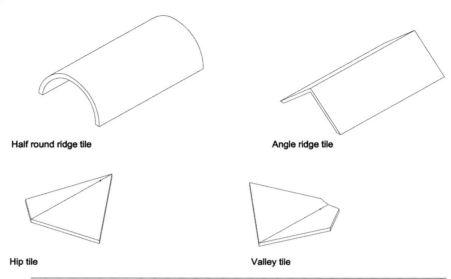

Half round ridge tile

Angle ridge tile

Hip tile

Valley tile

Figure 14.25 Special tiles

Special tiles

Purpose made tiles are produced for features such as ridges, hips and valleys (see Figure 14.25). These are normally manufactured from either clay or concrete.

At the verge and ridge the tiles are bedded in 1:3 cement:sand mortar. At the verge an *undercloak* should be provided bedded directly onto the brickwork of the gable wall. Alternatively proprietary dry verge and ridge systems are now available using mechanical fixing systems rather than relying on cement:sand mortar.

On natural slate roofs, hips, valleys and ridges must either be covered with clay or concrete tiles or these features may be covered with a malleable sheet metal such as lead or copper.

Self-assessment question 14.11

Why cannot natural slate tiles be used to form the hips, valleys and ridges on a slate roof?

An alternative to valley tiles on clay or concrete tiled roofs is to provide a valley gutter. This can be fashioned from lead or copper sheet or a pre-formed gutter manufactured from glass reinforced plastic may be used.

Flashings

At vertical abutments, where the roof meets projecting external walls or where a chimney stack or soil vent pipe protrudes through the roof, the gap created between these features and the roof covering must be covered with an impermeable material. The brickwork in any projecting wall or chimney stack must also be protected from becoming damp from the rainwater draining from the roof. At these points a *flashing* must be provided, cloaked over or underneath the tiles and dressed a minimum of 150mm up the abutment wall or feature and turned into a brickwork joint (see Figure 14.26). A similar flashing is used for projecting soil vent pipes in the shape of a cloak.

Because of the slope of the roof the flashing needs to be stepped to maintain the minimum height of 150mm above the surface of the roof. At the rear of a chimney the rainwater draining down the roof will need to be conducted around the rear of the chimney. At this point a gutter needs to be fashioned from timber or plywood boarding and covered with the material used for the flashing (see Figure 14.26).

Flashings are normally fabricated from sheet lead, copper or aluminium, although synthetic materials are available for pre-formed cloaks. Lead flashings are normally of code nos. 3, 4 or 5.

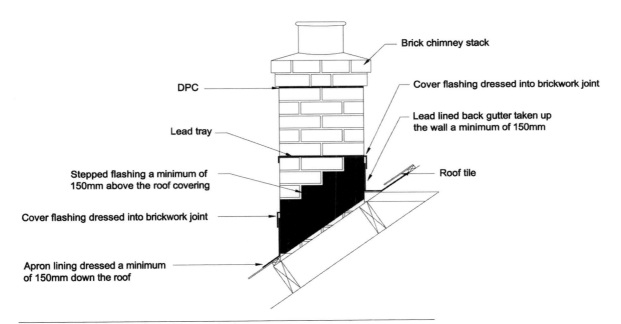

Figure 14.26 A flashing to a chimney stack

 Visit the companion website to test your understanding of Chapter 14 with a multiple choice questionnaire, and to see 59 full-colour photos illustrating this chapter.

CHAPTER 15

Timber frame and steel frame house construction

Introduction

Timber frame is a form of construction in which the structural framing members of the building are fabricated from timber and connected together in order to transmit the building loads to a suitable foundation onto which they are based. The framework forms the main shell of the building and suitable external claddings are subsequently attached to provide the necessary weather protection. The framework may be constructed up to three storeys in height, although 5-storey buildings have now been constructed using timber frame technology. Steel frame housing is similar in concept where, instead of a framework of timber members, light steel sections are used to form the framework.

Many people consider timber frame construction as a relatively new form of constructing buildings, but in fact this method of construction pre-dates conventional masonry construction in many parts of the world and there are still good examples of timber frame buildings dating from the sixteenth century still in occupation in the United Kingdom. What is new about modern timber frame construction is the use of prefabrication to construct the frames within a factory and then transport them to site for erection on prepared bases. Steel frame houses are a relatively new concept in house construction in the UK, although they have been used for some years in North America, South Africa, Australia and New Zealand.

There are a number of advantages and disadvantages of timber frame and steel frame construction compared with conventional masonry construction. The advantages include:

- the reduced cost of financing the project due to:
  - swifter construction due to the use of prefabricated frames, planned component deliveries reducing the cost of having to buy materials long before they are required to be used and a consequent reduction of storage of materials on site;

- • fewer men and machines required to erect the houses on site, as much of the work has been undertaken in the factory;
- • fewer activities to co-ordinate on site making construction easier;

Self-assessment question 15.1

Why are there fewer activities to co-ordinate on site with timber frame and steel frame construction compared to traditional masonry construction?

- • a variety of designs are able to be produced at very little extra production cost due to the use of mechanised production systems and economies of scale in the factory;

Self-assessment question 15.2

Why should a wider variety of designs available be considered an advantage for this form of construction?

- • the prefabricated units may have services built in to them in the factory further reducing construction time on site;
- • timber and light steel sections, being lighter in weight than bricks and blocks, reduce the overall dead weight of the building, providing possible savings in foundation costs;
- • insulated timber frame and steel frame walls have much higher thermal properties than walls constructed from bricks and blocks;
- • timber frame and steel frame construction use dry construction, therefore reducing drying out time and allowing the building to be occupied much more quickly;
- • because the components are prefabricated in a factory, wastage and theft of materials on site is reduced.

There are also a number of disadvantages to timber frame and steel frame construction. These include:

- • Care is needed in the design and site erection of components since the structural integrity of the building relies on the way in which the framework is designed, fabricated and finally erected.
- • Care also needs to be taken in the design and site erection of components to provide adequate fire resistance.
- • Care needs to be taken in the construction of timber frame and steel frame buildings in order to reduce the occurrence of interstitial condensation within the external framework.
- • Timber frame and steel frame buildings tend to have poorer sound insulation performance than buildings constructed using traditional masonry walls. This problem can be overcome with adaptations to the method of construction.

Self-assessment question 15.3

Why might timber frame and steel frame houses have poorer sound insulation performance than houses constructed using traditional masonry walls?

- Measures need to be taken to accommodate differential movement between the timber frame or steel frame structure and the external cladding.

Self-assessment question 15.4

Why should differential movement between the timber frame or steel frame structure and the external cladding be a problem?

- Timber frame and light steel frame houses have a lower thermal mass than equivalent houses of masonry construction.

Self-assessment question 15.5

Why should having a lower thermal mass than masonry construction be seen as a disadvantage for timber frame and steel frame houses?

- There is still an amount of consumer reluctance to accept any non-traditional forms of construction. This has been fuelled by reports in the media of poor construction practices associated with timber frame construction and the unfounded concern shown by potential buyers of timber framed houses that mortgage lenders may be reluctant in the future to advance mortgages on properties built using timber frame or steel frame construction.

There is no doubt that timber frame construction has sporadically had a bad press over the past 30 years or so. In fact the devastating impact that the *Man Alive* programme in 1984 had on this sector of construction activity, can be noted from the drop in the number of houses constructed using this technology from approximately 50 per cent of all new house starts before the programme was aired, to approximately 5 per cent of all new house starts in England and Wales, following the broadcast. The house-building sector of the construction industry in the United Kingdom has learned a lot from this criticism and in recent years the quality of timber frame house construction has improved and its market share has gradually risen.

This chapter will consider the main methods of timber frame and steel frame house construction used in the United Kingdom and the methods of providing an external cladding to these buildings. Consideration is then given to the measures needed to address the main concerns with

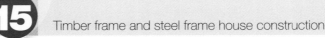

these forms of construction; those of structural stability, precautions against decay and corrosion, fire resistance and sound insulation.

Methods of timber frame construction

There are basically four main methods of timber frame house construction:

- platform frame
- balloon frame
- post and beam
- volumetric units.

Platform frame

This is the most commonly used form of timber frame construction in the United Kingdom. All timber panels are storey height. Ground floor to first floor panels support first floor joists, and first floor to second floor panels support roof trusses in a conventional 2-storey structure. The first-to second-storey panels are erected on top of the first floor platform, hence the name platform frame (see Figure 15.1).

This form of construction allows easily manageable storey height panels to be prefabricated off site, transported to site and rapidly erected on site, without the need for heavy lifting equipment. The designer also

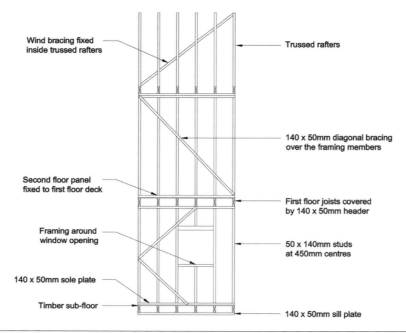

Figure 15.1 Platform frame construction

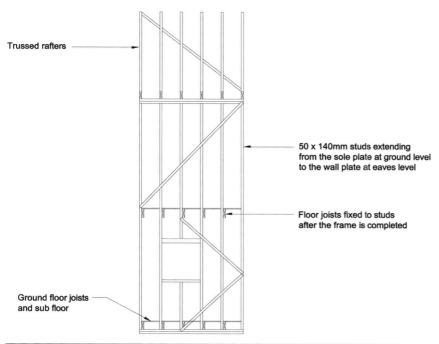

Trussed rafters

50 x 140mm studs extending
from the sole plate at ground level
to the wall plate at eaves level

Floor joists fixed to studs
after the frame is completed

Ground floor joists
and sub floor

Figure 15.2 Balloon frame construction

has complete freedom of interior room layout, as most of the internal walls will be non-loadbearing, along with the positioning of the doors and windows in the external panels.

Balloon frame

The external timber frames extend from the *sole plate* at ground level to the eaves. These frames then support the joists for the upper floors and the trussed rafters for the roof. The internal walls are inserted following the completion of the external structure (see Figure 15.2).

This form of construction is seldom used in the United Kingdom, as the lengths of timber required are difficult to obtain. Also lifting equipment is generally required on site in order to erect the large and heavy external frames into position. This form of timber frame construction has, however, been popular in Scandinavia and North America, where longer lengths of timber for these frames are more readily available.

Self-assessment question 15.6

What could be considered to be an advantage of balloon frame construction compared with platform frame construction?

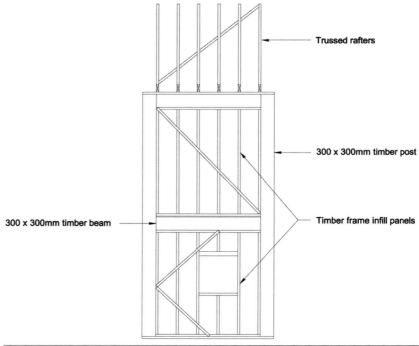

Trussed rafters

300 x 300mm timber post

Timber frame infill panels

300 x 300mm timber beam

Figure 15.3 Post and beam construction

Post and beam

A structural grid framework is produced using large section timber posts, supporting large section timber beams. These support the upper floor joists and roof trusses (see Figure 15.3).

The external infill panels and internal partition walls are similar to those used in platform frame construction. Thus the structural frame becomes superfluous on short span structures, such as most houses.

Self-assessment question 15.7

Why is the post and beam framework in this type of timber frame construction considered to be superfluous on short span structures?

This form of construction is similar to that used on historic timber framed buildings and tends to be chosen by designers who wish to emulate the exposed timber framework that is a characteristic of these historic buildings.

Volumetric units

These are prefabricated 'box' units constructed in the factory from floor, wall and roof panels and incorporating, where desired, all internal

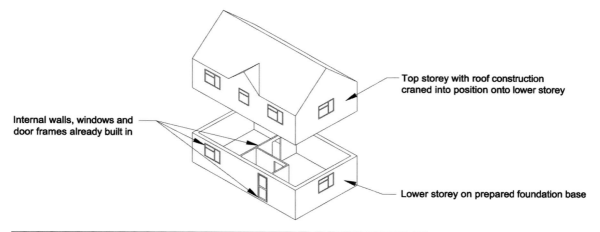

Top storey with roof construction craned into position onto lower storey

Internal walls, windows and door frames already built in

Lower storey on prepared foundation base

Figure 15.4 Volumetric units

services and finishes. The completed storey height units are transported to site, lifted into position by a crane and joined together to form the house (see Figure 15.4). They merely require the addition of external cladding materials to finish off the building.

This form of construction tends to be restricted to smaller properties, due to the limitations of transportation and lifting. They are also not favoured by some architects and planners, due to their repetitive design. However, they can be constructed extremely quickly and are therefore proving ideal for the construction of low cost, affordable housing.

Structural stability of timber frame

Timber frame structures for houses are covered by *BS 5268–6.1:1996 Structural use of timber. Code of practice for timber frame walls. Dwellings not exceeding seven storeys*. The timber used must be selected according to species, grade, strength, size and moisture content.

The frame is normally based on a 140 x 50mm timber *sole plate* (sometimes called a *base plate)* fixed to the solid ground floor slab. The sole plate must be accurately set out and levelled or the whole of the frame will be out of alignment and may be unstable.

Wall panels are fabricated from 140 x 50mm vertical timber *studs* spaced at 400–600mm centres and fixed between 140 x 50mm horizontal timber *head plates* and *bottom plates* or *sill plates*. The frame is prevented from *racking* or dropping out of shape by means of 140 x 50mm timber diagonal *braces* extending across the framework (see Figures 15.1 and 15.2). Fixing plywood, oriented strand board or cement-based particleboard sheets as *sheathing* to the exterior provides additional support to the framework (see Figure 15.5). This is covered by *BS EN 594:1996 Timber frame structures. Racking strength and stiffness of timber frame wall panels*.

The floor decking can be constructed of tongued and grooved plywood sheets or sheets of particleboard, oriented strand board or flaxboard. The pitched roofs are normally constructed from trussed rafters.

Timber framed houses may use suspended timber ground floors. Where these are used the sole plates are fixed on top of the sleeper walls.

External wall finishes for timber frame

Timber framed buildings require an external wall finish to provide adequate weather resistance. Suitable finishes include:

- brick or masonry
- timber cladding, either horizontal or vertical
- tile hanging
- cement and sand rendering.

These finishes are secured to the sheathing, allowing differential movement to occur without damaging the finish or the timber frame. A special flexible wall tie has been developed to fix brickwork external cladding to the timber frame (see Figure 15.5).

The use of weatherboarding, tile hanging and cement and sand rendering for external wall finishes is covered in more detail in Chapter 22 (external wall finishes).

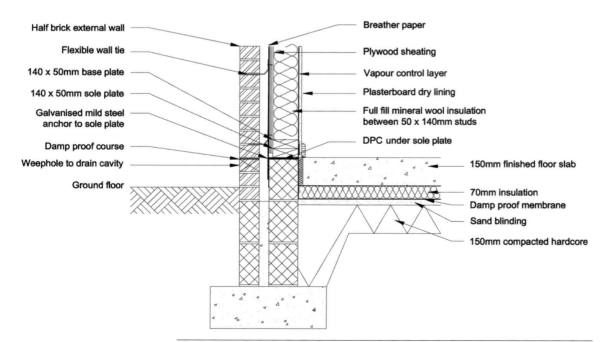

Figure 15.5 External wall finish to a timber frame

Precautions against decay

Timbers in direct contact with the ground or the external climate should be treated against decay. The sole plate should be bedded on a DPC (see Figure 15.5).

The timber used in timber frame construction should be delivered to site with a moisture content not greater than 20 per cent. When installed into a dry and heated building it should reach an equilibrium moisture content of between 12–14 per cent. This could result in a drying shrinkage of up to 10mm over a 2-storey structure. Clearly it is undesirable for the moisture content of the timber in the prefabricated frames to increase above that at delivery. The frames therefore need to be stored in dry conditions prior to erection. With *just in time* delivery management it should be possible to ensure that the frames are delivered just before they are erected, so reducing the need for on-site storage in dry conditions to a great extent.

Building Regulations Approved Document C, Section 5.17 stipulates that the cladding should be separated from the insulation by a vented and drained cavity with a membrane that allows the passage of water vapour out of the building, but resists the passage of liquid water into the building. This 50mm ventilated cavity allows for rapid drying of any water that may penetrate the exterior cladding and also enables water vapour escaping from the building to condense into water droplets and be drained or ventilated away.

The *breather membrane* is applied to the exterior surface of the sheathing (see Figure 15.5). This generally consists of a sandwich membrane comprising a high density polyethylene outer layer encasing a micro-porous film. The membrane is moisture resistant but water vapour permeable. Thus it prevents rain penetration from outside, that may have got past the external cladding getting into the building, but enables any water vapour that may be passing through the wall construction from inside the building to percolate through to the external environment.

Timber framed houses have high thermal insulation values. In consequence, excessive water vapour produced in the house will percolate through the wall structure and may condense on cold surfaces within the wall framework, causing interstitial condensation. This is a similar problem to that encountered in roof construction and similar precautionary measures should be used to overcome it. As it is difficult to effectively ventilate the wall, a vapour control layer should be applied on the warm side of the insulation. Polyethylene sheet is often used for this purpose. It should be lapped at its joints and sealed with tape. It is stapled to the vertical studs of the timber framework. Because of this it cannot be considered to be completely impervious. It is imperative that the vapour control layer remains intact and is not split. This can be a problem when trying to bring electrical cables from the cavity within the framework to sockets and switches that are positioned on the interior face of the wall. Any holes in the vapour control layer must be repaired with tape before the internal finish is fixed. A vapour control layer with a reflective surface is now being produced. This requires a gap to be created between it and the plasterboard internal lining. Heat escaping from the room can then be reflected back in. It is also claimed that this reflective layer can reduce solar heat gain through the wall structure from the exterior.

Thermal insulation

Either rigid thermal insulation boards or a combination of rigid boards and flexible thermal insulation materials can be placed between the vertical stud members and thus sandwiched between the plywood sheathing on the outside of the timber frame and the plasterboard internal finish on the inside of the frame (see Figure 15.5). This enables thermal insulation up to 140mm in thickness to be installed in the timber frame wall and U-values as low as $0.27W/m^2K$ to be obtained. This is often better than the thermal performance that can normally be achieved by a traditional brick and block cavity wall with thermal insulation contained within the cavity, and meets the maximum area rated U-value for walls of $0.35W/m^2K$ laid down in Building Regulations Approved Document L1A. Further improvements to this U-value can be achieved if plasterboard with enhanced thermal insulation properties is used.

Timber frame construction is more effective at reducing air permeability than masonry construction. This is because the joints of the plasterboard dry lining, the vapour control layer and the breather membrane can be effectively sealed to keep any air leakage to a minimum.

Fire resistance

The fire resistance of the roof, floors, non-loadbearing internal walls and stairs of timber framed houses is similar to that of the equivalent elements in traditional house construction. Plasterboard *dry lining* to walls and ceilings provides acceptable fire resistance in compliance with Building Regulations Approved Document B, section B4. Dry lining is discussed in more detail in Chapter 21 (internal finishes). The fire resistance of the timber can be improved by treating it with a fire retardant solution by pressure impregnation in a similar manner to that used to provide preservative treatment.

To prevent fire from spreading within the cavities of a timber framed house, cavity barriers in the form of incombustible quilt are provided around all openings, horizontally at joist and eaves levels and vertically at intervals to break up large areas of the cavity. Building Regulations Approved Document B, Section 6.5 recommends that the cavity barriers should be at least 38mm thick in timber construction in order to provide at least 30 minutes fire resistance.

Sound insulation

Insulation against airborne sound transmission relies largely on the mass of the structure, particularly the external walls. Timber frame houses having less mass than traditional masonry construction must therefore rely to a large extent upon separation between elements of the structure to achieve sound reduction, rather than attempting to incorporate more mass into the structure. The design of external walls in timber frame construction where the cladding is separated from the frame will generally be sufficient to satisfy this requirement. Where a separating party wall needs to be constructed between adjoining properties, Building Regulations Approved Document E, Section 2.147 recommends that there should be a minimum distance of 200mm between the inside faces of the linings. The linings should comprise two or more sheets of plasterboard, each sheet having a minimum mass of 10kg/m^2 and having staggered joints. Sound absorbent mineral wool batts or quilt should be placed in the cavity between the frames. This should have a minimum thickness of 25mm if suspended between the frames or 50mm if attached to one of the frames.

Sustainability

As discussed in Chapter 10 (timber), wood is a natural organic material which is non-toxic and has the benefit of being a renewable resource. Indeed the amount of softwood timber being grown in Northern Europe far outweighs the amount used by the construction industry. Timber is also recyclable and biodegradable and, perhaps most importantly, is carbon neutral in that it takes up CO_2 during its growth and stores it until it may be released by incineration. It is an excellent insulating material, having a low coefficient of thermal conductivity.

It does have some embodied energy from the conversion process, kiln seasoning (if used) and preservative application. It also uses energy in transportation and this may be high where imported timbers are being used. However, it has been estimated that the embodied energy in the materials used to construct a typical three bedroom house is approximately 20 tonnes where conventional masonry construction is used, but can be as low as 2.4 tonnes where timber frame construction is used with timber also being used for the external cladding, the floors, doors and windows of the house.

Steel frame house construction

Currently light steel frame house construction is not as popular as timber frame house construction, but it has similar advantages to the timber system, as well as being lighter in weight and not having the same problems with moisture movement as the timber system has. Light steel frames also exhibit similar performance to timber frames regarding enhanced resistance to air permeability. The disadvantages of the two systems are very similar. Interstitial condensation occurring in the steel frame system could lead to corrosion of the framing members. Light steel frames also have similar problems with fire resistance and sound insulation to those experienced by timber frame houses. There is also concern about thermal bridging in light steel frames. If the insulation is placed outside the light steel frame sections then the methodology for calculating the U-value of the construction outlined in *BS EN ISO 6946:2007 Building components and building elements. Thermal resistance and thermal transmittance. Calculation method* can be adopted. Conversely, where the steel framing bridges some or all of the insulation the above method for measuring the U-value of the construction cannot be adopted. However *Building Research Establishment Digest DG465:2002 U-values for light steel frame construction* provides an adaptation of the method used in *BS EN ISO 6946* that is suitable for determining the U-value of such constructions.

The sustainability of light steel frame construction is not as good as timber, due to the amount of embodied energy required to produce the steel and to press the sections. In addition, the sections used in light steel frames have high thermal conductivity, which can cause problems with thermal bridging. However, the steel members are much lighter than the comparable timber members, so energy used in transportation is less. Also steel is highly recyclable. Where the frames are constructed in the factory, all cut-offs and even the swarf from drilling are put back into making new components.

The technology used in timber frame construction has also been adapted for light steel frame house construction and uses similar techniques to those used in the construction of steel frames for lightweight partitions (see Chapter 20 on internal walls and partitions). There are a few proprietary systems that have been developed. A typical system utilises 1.6mm thick galvanised, cold-rolled steel strips that have been formed into 100 x 50mm C section studs and 100 x 50mm channel sections. These sections can be supplied to site as individual components for assembly in situ, either cut to the required length or in standard lengths for cutting to suit on site. The channel sections, known as tracks, are positioned at the base and head of the frame, performing a similar function to the sole plate and head plate in a timber frame. The C sections are then fitted vertically between the tracks at 600mm centres, performing a similar function to the vertical studs in timber frames. Diagonal bracing to prevent racking of the frame is provided by 90 x 1.2mm flat plates. All members are fixed by self-tapping screws. The frames may use either platform frame or balloon frame formats, although platform frame is the preferred format in the United Kingdom.

Light steel frames can also be assembled using prefabricated 1200mm wide wall panels comprising vertical C section studs spot welded in the factory to horizontal top and bottom channel section tracks. Additional diagonal braces, in the form of flat plates, are added for stability and to strengthen the frame to resist wind loading. The ground floor to first floor panels are fixed to a galvanised steel base track, acting as a ring beam, that has been concreted to the ground floor slab. Upper floors may be constructed from 200mm deep light steel C section joists at 600mm centres supporting a floating timber floor comprising 15mm ply sheeting with 19mm plasterboard topping supporting 53mm timber battens at 600mm centres with 18mm tongued and grooved particleboard decking. Alternatively an in situ concrete floor slab can be cast or a precast concrete beam and block floor can be used. However, concrete floors are more likely to be used on multi-storey apartment blocks than individual houses. The roof can be constructed from light steel C section members and assembled in situ in a similar manner to that of timber cut roof construction, but commonly the pitched roof is constructed from timber trussed rafters.

The external walls of the frame can be clad with a half brick thick wall, timber boarding or render. Where brickwork is used, narrow steel channels are fixed to the outside of the wall frames to accommodate 'fish tail'-shaped brick ties that are embedded into the joints of the brickwork cladding. This enables any differential movement between the frame and the cladding to take place (see Figure 15.6). Where a rendered finish is to be used a 12mm sheathing board is fixed to the studs of the framework and a light steel rail is fixed through this to the studs to support the render and provide the 50mm drained and ventilated cavity. Similarly, where timber boarding is to be used the boards are supported by 50mm timber battens fixed through the sheathing to the vertical studs in the framework. The inner faces of the external wall panels are covered with two layers of plasterboard dry lining.

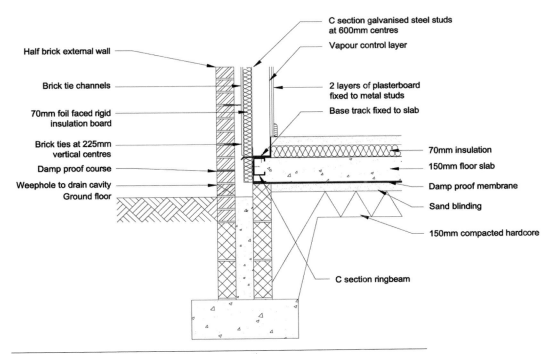

Figure 15.6 Typical steel frame house construction

Unlike timber frame construction, light steel frame systems are of bespoke design from architect's drawings supplied to the manufacturers and are erected by using specially trained erection teams selected by the manufacturers.

 Visit the companion website to test your understanding of Chapter 15 with a multiple choice questionnaire.

C H A P T E R **16**

Windows and glazing

Introduction

The primary function of a window is to allow daylight into the building. Before the invention of glass, windows were open apertures in the external walls of buildings and allowed wind and rain to enter through them. In consequence, the apertures were small and not particularly efficient at allowing daylight to enter the building. Modern windows comprise a sheet (or sheets) of transparent or opaque glass enclosed in a framework of timber, steel, aluminium or PVCu (plastic). This enables windows to be much larger and their daylighting efficiency is much improved.

The Building Regulations do not specify minimum daylight requirements for a building. However, if the window area is small there will be an increased dependency on internal lighting, thus increasing the amount of energy used by the building with its consequent increase in the CO_2 emissions generated to produce the extra electricity being used. It is therefore recommended that the minimum area of glazing used in a building should be 20 per cent of the floor area.

Windows can have the sheet of glass fixed in an enclosing framework or the window frame can accommodate smaller sub-frames or *casements* that can be opened. This enables the window not only to allow daylight to enter the building but also fresh air from the exterior to ventilate the building. Modern windows may have a number of differing sized casements in one frame and this allows varying amounts of ventilation to be accommodated.

If windows contain transparent glazing then they may serve an additional function in providing acceptable viewing of the external environment to occupants of buildings. This may have a psychological benefit where occupants of the building are likely to spend a large part of each day separated from the outside environment. Designers will therefore often consider placing the windows in strategic locations to afford acceptable viewing to the building occupants.

The location of windows in the façade of the building is also of importance to the appearance of the building. Indeed the style of window used in a building often has a strong influence on the building's architectural style (consider the influence of the style of windows used in Gothic, Georgian and Art Deco styled buildings). This aspect of architectural styling and the disposition of windows within the façade of buildings is termed *fenestration*.

It is possible to manufacture windows on site, but it is generally considered to be more economical to manufacture them in factories and deliver them to site for installation into the external walls of the building. Apart from the wide range of materials and styles that windows can be manufactured from, there is also a wide range of window manufacturers. In order to ensure that windows from different manufacturers are of similar sizes there is a need for a system of *dimensional co-ordination*. This enables manufacturers to offer a variety of standard sizes that will suit the architect's desire to provide a certain amount of flair in the design, but also ensure that the amount of variability is not too great to detrimentally affect the economies of scale brought about by the mass production of components. Variations in size are therefore based on increments of a standard unit measurement or *module*.

With the increasing use of prefabrication for the manufacture of components for buildings it has become necessary for this standard unit of measurement to be adopted by all manufacturers of all building components in a system of *modular co-ordination*. Unfortunately it has not been possible to agree on the standard size of a module for all building components, but there is general agreement amongst manufacturers of similar components regarding the basic module size for their component. Window manufacturers have tended to base their basic module size on the dimensions of the standard brick.

Self-assessment question 16.1

Why should window manufacturers have based their basic module size on the dimensions of the standard brick?

This chapter will first consider the functional requirements of windows and then compare and contrast the four main types of materials used in window manufacture. The design of a variety of different types of window will then be discussed. The glazing and ironmongery used for windows will also be considered. Finally, consideration will be given to the methods of installing windows into houses.

Functional requirements

Apart from the primary functional requirement and supplementary functional requirements already considered in the introduction, windows will also need to satisfy the following additional functional requirements.

Weather resistance

Because windows form part of the external fabric of the building, they must be able to offer a similar level of weather resistance to that of the walls in which they are situated. Fortunately much of the window consists of glass, which is an impermeable material. However, where the window design incorporates opening casements it is necessary to ensure that rain penetration and air infiltration does not enter the building. This can be difficult to achieve, as there needs to be a sufficient gap between the casement frame and the main window frame that the casement is fitted within to enable the casement to open and close satisfactorily. In consequence, a lot of thought has gone into the detailing of this joint to reduce the problem of rain penetration and air infiltration. Consideration also needs to be given to the detailing of the joint between the main window frame and the head, cills and reveals of the wall opening into which it is fixed. Both these aspects are considered in more detail later in this chapter.

Security

Windows can provide a means of access into the building, particularly where opening casements are included in their design. It is therefore important to ensure that good quality catches and locks are used on opening casements to prevent unwanted intruders gaining access to the house. It is also important to consider ways of preventing the panes of glass from being removed from the outside of the window, enabling undesired access into the building.

Thermal insulation

As windows form part of the external fabric of the building, their thermal performance will affect the overall thermal performance of the external wall in which they are situated. Unfortunately much of the window is composed of glass, which has a high thermal conductivity. It is estimated that a typical house can lose up to 10 per cent of its heat through the windows. The thermal performance of windows can be improved by the use of multiple glazing. This consists of two or more panes of glass entrapping an air space. There are four main types of multiple glazing:

- Double glazed windows – comprising factory sealed panes of glass fixed into a wider frame rebate than used in single glazing (see Figure 16.1). This method is the most popular for modern windows.
- Glazed in situ double glazing – comprising two panes of glass fixed in situ into a double rebated frame (see Figure 16.2) or contained in separate beading fixed to the existing frame (see Figure 16.3).

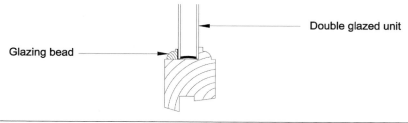

Figure 16.1 A double glazed window

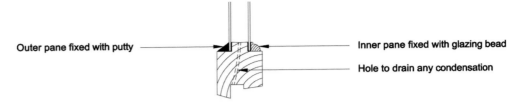

Outer pane fixed with putty ——————————— Inner pane fixed with glazing bead

———————— Hole to drain any condensation

Figure 16.2 Two panes of glass fitted into a double rebated frame

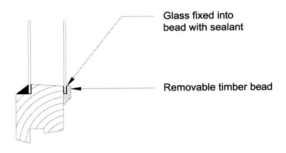

Glass fixed into bead with sealant

Removable timber bead

Figure 16.3 Two panes of glass retained in beading fixed to the frame

- Coupled windows – comprising two panes of glass fixed in separate casements and coupled together (see Figure 16.4).
- Double windows – comprising a secondary system of glazing contained within its own frame and fixed to the inside of the existing single glazed window frame (see Figure 16.5). This form of multiple glazing has been popular in retrofit applications where existing single

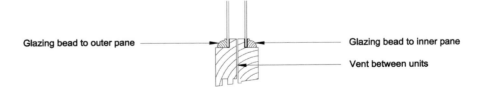

Glazing bead to outer pane ——————————— Glazing bead to inner pane

———————— Vent between units

Figure 16.4 A coupled window

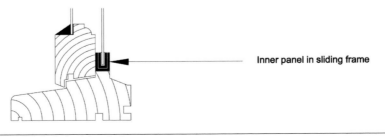

Inner panel in sliding frame

Figure 16.5 A double window

glazed windows have been upgraded to multiple glazing without removing the existing window unit.

The loss of heat through a window can occur through a combination of radiation, convection and conduction. Nearly 66 per cent of the energy lost from a window is from radiation through the glazing. In a double glazed unit the inside pane will absorb heat from the room and then transmit this through a combination of convection and conduction across the cavity between the two panes of glass to the cooler outside pane and from there to the outside. The gauge used to measure the heat energy emitted by the surface of the glass is its *emissivity*.

A small amount of heat is lost through convection within the glazing cavity between the two panes of glass. Air within the cavity is warmed by the inner pane of glass. This warm air then rises and is replaced by cooler air, setting up a convection current which transfers the heat to the outer pane of glass. Heat losses through convection can be reduced by sealing inert gases such as argon, krypton or xenon, which are heavier than air, into the glazing cavity. The integrity of this fill is only as good as the effectiveness of the seal between the panes making up the glazing unit. It has also been found that if the cavity between the panes of glass is restricted to below 20mm, losses through convection are minimised.

Heat losses through the glazing element of the window are more significant than those through the frame element, as the thermal transmittance coefficient of glass is much greater than that of the material used for the frame (particularly if the frame is timber). Also the surface area of the glazing in a window is far larger than the surface area of the frame. However, heat can be conducted through the frame of the window. The rate of this conduction is related to the material used for the frame. In general, frames made from timber conduct less heat than those made from steel or aluminium. These materials also suffered in the past from problems of surface condensation on the frame as well as the glazing. Modern aluminium windows now incorporate thermal insulation materials within the hollow extrusions of the frame, and along with the use of multiple glazing these windows have now considerably reduced the thermal bridging effect previously suffered. Some conduction heat loss also occurs through the spacer bars used to separate and seal the panes of glass in a glazing unit. These are usually made from aluminium, but other non-metallic materials such as glass fibre and structural foams are now also being used to reduce the effect of thermal bridging. The U-value of a window should include both the frame and the glazing unit. The Building Regulations Approved Document L1A Section 4.21 sets out limiting U-values for various elements of the building. These are weighted average values (as explained in Chapter 1) and for windows this value is 2.00W/m^{2}K.

Self-assessment question 16.2

Why do timber frames have better thermal insulation values than metal frames?

Heat losses can also occur through the gaps between the opening casements and the window frame. These heat losses can be reduced by the inclusion of *weatherstripping* to the casement frame. Weatherstripping normally consists of a compressible material that can effectively seal the gap between the casement frame and the window frame when the casement is closed. Not surprisingly, opening windows suffer higher air permeability than fixed windows, and casement windows generally perform better than traditional sash windows. Air permeability can also occur between the frame and the wall opening. Gaps between the frame and the wall should be sealed with low expanding foam, backed up with a single-sided, pre-folded, high performance adhesive tape to provide an air tight bond between the window frame and the wall.

Although not directly related to thermal insulation the topic of comfort is also an important consideration. It has been found that the occupants of a room will feel discomfort if the temperature of a surface within the building fabric is 4^{0}C lower than the air temperature within the room. This discomfort is generally alleviated by increasing the room temperature by a few degrees. Where the external walls of the house are well insulated the temperature difference between the internal surface of the wall and the air within the room is minimal, but the temperature difference between the internal surface of the glazing and the air within the room can be large enough to cause discomfort, particularly if the window has only single glazing, where the temperature difference can be as high as 20^{0}C. This difference can be reduced to 8^{0}C where double glazing is used and 4^{0}C where triple glazing is used.

Windows not only lose heat to the external environment, they can also absorb heat from the sun (known as solar gain). This can occur through two mechanisms: primary transmittance by radiation through the glazing from the exterior to the interior of the building, and secondary transmittance through convection and conduction caused by the solar energy heating up the outside pane of glass and the heat being transferred across the glazing cavity to the inside pane of glass and subsequently to the interior of the building. This solar gain can make buildings uncomfortable to use in the summer months. Glazing is now assigned a G-value. This measures the amount the glazing blocks heat from the sun. The lower the glazing unit's G-value the less solar heat it transmits. The amount of solar energy gained by the glazing will also be dependent on factors such as orientation (the direction the window faces) and shading.

Sound insulation

Similarly to thermal insulation the overall sound insulation of the external wall will be affected by the sound insulation of the window units located within it. Once again the use of multiple glazing can make a significant difference to the performance of the window. *Building Research Establishment Digest 337: Sound insulation: basic principles* states that, unlike the gap between panes for optimum thermal performance, the gap between panes for optimum acoustic performance is considerably wider;

200mm wide for glass thicknesses of between 3–4mm, 150mm wide for glass thicknesses of between 4–6mm, and 100mm wide for glass thicknesses greater than 6mm. This suggests that it is not possible to satisfy the requirements of effective thermal insulation and sound insulation with the same window. In fact some acoustic attenuation is achievable with a 19mm gap between the panes in multiple glazing and most homebuyers are likely to seek high thermal performance at the expense of less enhanced acoustic performance from the windows in their new house.

Weatherstripping around the edges of opening casements also helps to reduce sound transmission through the gap between the casements and the window frame. Any gaps between the frame and the external wall should also be well sealed, as discussed previously, to prevent sound transmission in these positions.

Cleaning

To provide optimum light transmission and to enable occupants to view the outside environment through the windows, the glazing needs to be frequently cleaned. This is generally not a problem in houses up to three storeys in height where the windows to rooms on the upper storeys can be reached by a ladder, but in taller buildings, such as in blocks of flats or where a window to an upper room cannot be easily reached because of a projection such as a conservatory on the ground storey, it may not be possible to reach the external face of the window to clean it. In these situations, windows that enable the outside face to be cleaned easily from the inside of the building, such as the pivot window, may be considered to be a better choice than the conventional side hung or top hung casement window.

Materials

Windows are generally manufactured from one of four main materials:

- timber
- steel
- aluminium
- PVCu (plastic).

Each material has its own advantages and disadvantages.

Timber

Timber is still the most popular material for windows, due to its availability, low cost, ease of working and appearance. Timber windows also use less energy in their production than steel, aluminium and PVCu windows and are manufactured from a renewable resource, which is good for sustainability. They are covered by *BS 644:2009 Wood windows. Fully finished factory-assembled windows of various types.*

Specification. Because wood is a good thermal insulator there is little problem with surface condensation on the frames of timber windows.

The disadvantages of timber windows are their high moisture movement, which can make casements difficult to open and close in winter when the timber has swollen. Any remedial treatment such as planing around the edge of the casement to ensure a better fit following moisture movement of the timber can result in an increased gap between the casement and the frame when the timber has shrunk in drier conditions. This increased gap leads to a greater amount of air infiltration and a reduced thermal and acoustic performance. It is for this reason that it can be difficult to weatherstrip timber windows. However, modern high performance timber windows are now supplied double glazed and weatherstripped. Timber windows are vulnerable to wet rot unless they are manufactured from hardwood or are adequately protected. They may be protected by painting or staining once every 3–4 years. Alternatively timber windows are available that have been clad in plastic or faced in aluminium at the factory.

Timber windows are normally fixed into position by the builder as the external wall is constructed. The glazing is normally fitted into position with linseed oil putty before the building is made weathertight. High performance timber windows are generally delivered with sealed double glazing units already fitted.

Steel

Hot rolled steel sections to *BS 6510:2010 Steel framed windows and glazed doors. Specification* are quite inexpensively produced and are easily assembled within the factory to produce window units. The sections used are small, providing optimum daylight admission.

The disadvantages of steel windows are that they are not seen as being as attractive as timber windows. They also suffer from higher thermal transmittance coefficients than timber windows. Staining of internal wall finishes by surface condensation may be alleviated by placing the steel windows in a timber sub-frame. Steel windows are prone to corrosion and so must be hot-dip galvanised. They should be further protected by painting once every 4–5 years. Alternatively they may be supplied with a polyester powder coating that is available in a range of colours. Stainless steels are available for window manufacture but tend to be too expensive for consideration in housing.

Steel windows are 'built in' as the external wall is constructed, in a similar manner to that of timber windows. They are normally glazed before the building is made weathertight using putty containing a special hardening agent.

Aluminium

These are constructed from hollow extruded sections of type 6063 aluminium alloy to *BS 4873:2009 Aluminium alloy windows and doorsets. Specification*. They are supplied in one of four types of finish. Mill finish is the natural finish. Anodised finish has a protective coating obtained from electrolytic oxidation. Organic finish is produced using acrylic or polyester powder coatings in a variety of colours. The aluminium sections may also be clad in stainless steel. All these finishes provide a high standard of durability. They also give aluminium windows a pleasing appearance.

The disadvantages of aluminium windows include their high initial cost (that may be offset by their low maintenance requirements and long life span) and their high thermal conductivity, which can cause surface condensation on the frames and also provide a cold bridge of heat flow from the interior to the exterior of the building. This has been quite effectively overcome by filling the hollow frame sections with thermal insulation material. Aluminium is a relatively soft metal and window frames can easily become scratched or dented during the construction process. They are therefore fitted at a later stage in the construction sequence than is normally the case with timber or steel framed windows.

Aluminium windows come to the site factory glazed and weatherstripped. The sealed double glazing units are installed in the frame using a *neoprene gasket*. Aluminium window manufacturers normally offer a supply and fit service, rather than supplying the window units for the builder to fit on site. Because they are installed at a late stage in the construction sequence they are fitted into prepared openings formed in the external walls by temporary timber templates.

PVCu

Unplasticised polyvinylchloride windows have only been used in the United Kingdom for approximately 40 years. Prior to this they were popular in continental Europe, particularly Germany. They are made from hollow extruded sections, similar in shape to the extruded aluminium sections previously considered, to *BS 7412:2007 Specification for windows and doorsets made from unplasticized polyvinylchloride (PVC-U) extruded hollow profiles*. However, they are much larger in size and therefore for the same area of window have the least area of glazing of all four materials considered. Thus the amount of daylight admitted by these windows is less than with the other framing materials. PVCu windows have good durability. There was concern when they were first introduced to the United Kingdom that they would suffer from ultra-violet degradation and become brittle when exposed for long periods of time to sunlight. This has now not been found to be a problem, but although they are available in a wide range of colours the brighter colours are probably best avoided, due to their tendency to fade when exposed to sunlight over a long period of time. Plastic window frames have a similar thermal conductance value to that of timber frames and do not suffer from the thermal bridging effect that can beset steel and aluminium windows.

The disadvantages with these windows include an initial consumer reluctance to accept them, since plastic had an association with 'cheap and shoddy' goods. In fact these windows are quite expensive to purchase and have a similar initial cost to aluminium windows. They do however compare favourably with aluminium windows with regard to good durability and low maintenance costs. The extruded hollow sections of PVCu windows are not as strong as the equivalent sections in aluminium windows. In consequence, on large windows the hollow sections may need to be reinforced with steel cores.

PVCu windows can be easily scratched or damaged during construction. In consequence they are generally fitted late on in the construction sequence, similar to aluminium windows. As with aluminium windows they are supplied ready glazed and weatherstripped and are fitted into prepared openings by the manufacturers' own fitting team.

Types of window

Windows are classified by their method of opening into ten main categories, as shown in Table 16.1.

Table 16.1 Window types

Type	Attributes	Design
Fixed	The cheapest form of window No opening facility	
Top hung casement	Usually small in size Usually situated at high level Provide limited ventilation	

(Continued)

Type	Attributes	Design
Bottom hung casement	Hopper type opening Opens inwards for some access to outer face Limited accessibility for intruders	
Side hung casement	Most popular design Large ventilation facility Vulnerable to distortion if too large	
Horizontal pivot	Particularly easy to clean from the inside of the building Weatherproofing can be difficult	
Vertical pivot	Similar to above	

Table 16.1 *(Continued)*

Type	Attributes	Design
Horizontal sliding	Fine dimensional tolerances Bottom tracks vulnerable to fouling Weatherproofing can be difficult	
Vertical sliding	Can be made very tall Good ventilation control Modern techniques for opening and closing	
Moving axis	Tilt and turn mechanism Provides easier access for cleaning	
Louvred	Rapid ventilation is possible Limited accessibility for intruders Poor weatherproofing	

Standard timber casement window

This style of window is available in a wide range of sizes and designs. The frame consists of a 70 x 45mm *head* member at the top and a 120 x 57mm *sill* member at the bottom, extending across the full width of the window, with 70 x 45mm *jamb* members at each side. The frame is further sub-divided by a 70 x 57 mm vertical *mullion* member and a 95 x 57mm horizontal *transom* member (see Figure 16.6). All these members are joined together with wedged *mortice and tenon joints*.

The frames have deep rebates to accommodate the opening casements, a small top hung *ventlight* and a larger side hung casement sash. A fixed light is accommodated beneath the ventlight. Each of these casements is formed by a sub-frame assembly comprising 45 x 40mm horizontal top and bottom rails and vertical stiles. These rails and stiles incorporate shallow rebates for glazing (see Figure 16.6).

Where the opening casements are accommodated in the members of the main frame, the gap between the casement and the frame needs to be wide enough to allow the casement to open and close without binding. In order to prevent rainwater from entering the building across this gap it is necessary to provide a groove or *throating* to the members to either side of the gap. This prevents the capillary attraction of water across the gap between the frame members (see Figure 16.6). This throating is not required to the framing for the fixed light as this will be

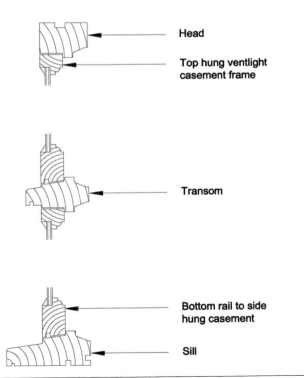

Head

Top hung ventlight casement frame

Transom

Bottom rail to side hung casement

Sill

Figure 16.6 A standard timber casement window

glued tightly into position and is not designed to be opened, thus a gap does not need to be created. The upper surfaces to the transom and sill members are sloping and a throating is created on their underside.

Self-assessment question 16.3

Why are the upper surfaces of the transom and sill members sloped?

The modified timber casement window

The standard timber casement window was found not to perform particularly well with regard to weather resistance and draught resistance. In consequence, the design of the timber casement window was modified in *BS 644:2009* in order to enhance its performance with regard to these two functional requirements.

The main design feature of this window is that the rails and stiles of the opening casements are rebated over the members of the main frame. This forms a double barrier to the entry of wind and rain (see Figure 16.7).

Where the window contains fixed lights the glazing can be fitted directly into the main frame members and unlike the standard timber casement window does not require a separate casement sub-frame (see Figure 16.7).

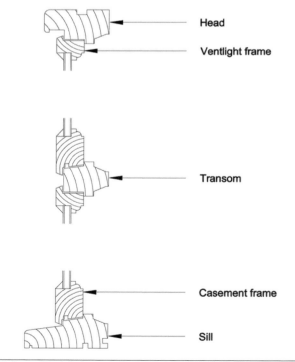

Head

Ventlight frame

Transom

Casement frame

Sill

Figure 16.7 A modified timber casement window

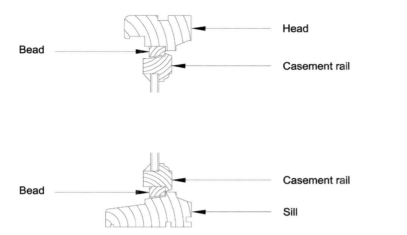

Figure 16.8 A timber horizontally pivoted window

The timber horizontally pivoted window

This window can be reversed by horizontally pivoting the casement about its centre so that the outside of the window can be cleaned from the inside.

The main problem with this window is that the rebates have to be reversed above and below the hinge and the weatherproofing associated with this requirement has to be carefully considered. Generally a rebated bead is fixed to the framing members above the hinge and to the casement members below the hinge (see Figure 16.8).

The other problem with this window design is that the hinge must be exposed on the outside of the window and will need to be made of corrosion resistant materials.

Self-assessment question 16.4

Why should the rebates to a horizontally pivoted window have to be reversed above and below the hinge?

The timber horizontally sliding window

These windows slide on bottom rollers fixed at sill level. A *water bar* is required behind the bottom roller to prevent the ingress of water and this may be masked by a cover strip.

At the head the sliding window is located in a guide rail that is installed in the lintel and a *weatherstrip* is located between the sliding casement and the frame of the fixed light that it slides over.

The timber vertically sliding window

These are usually based on a double hung sash in a cased frame. Traditionally, the vertical sashes were opened and closed by means of

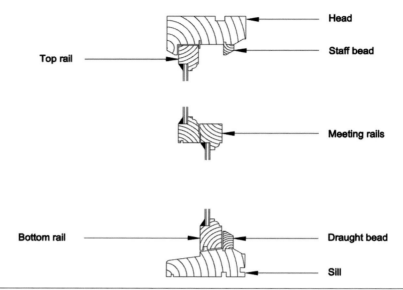

Figure 16.9 A timber vertically sliding window

weights hung on sash cords located within the cased frame. These tended to be bulky and cumbersome and once the cord carrying the weights had broken, the casing to the window needed to be removed and the whole mechanism renewed.

Modern designs now use spiral sash balances. These fit into a groove in the sash or the frame and work by means of a variable pitch spirally twisted rod revolving through a slotted bushing in order to wind and unwind a coiled spring inside the sash tube.

The frame consists of a 125 x 38mm head member, a 150 x 63mm sill member and 125 x 38mm stiles. The sashes comprise 50 x 44mm rails and stiles. A 19 x 19mm staff bead is fixed to the inside of the head member and a 38 x 22mm draught bead is fixed to the inside of the sill member. Also, 25 x 16mm parting beads are fixed to the head and stiles of the main frame to keep the two vertical sashes apart. The bottom rail of the top sash and the top rail of the bottom sash are rebated to form a weather seal when the sashes are in the closed position (see Figure 16.9).

Steel casement windows

All steel window sections are manufactured from rolled steel sections that are mitred and welded at their corners to form frames to *BS 6510:2010.* The standard 'Z' section is 25 x 33mm and can form the head, sill and jamb members of the main frame and the rails and stiles members of the casement sub-frames. The mullion section is a 25 x 45mm elongated 'Z' section and the transom is a 25 x 28mm 'T' bar (see Figure 16.10).

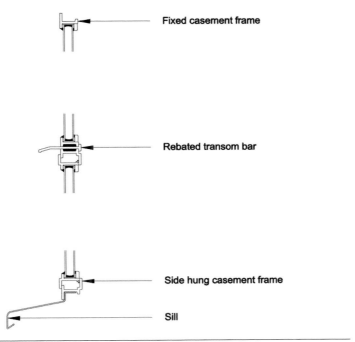

Fixed casement frame

Rebated transom bar

Side hung casement frame

Sill

Figure 16.10 A steel casement window

Aluminium windows

These are constructed from extruded HE9 aluminium alloy sections to *BS 4873:2009*. The sections are mechanically jointed or welded together into standard modular sized units (see Figure 16.11).

As mentioned previously, most aluminium windows now incorporate a thermal break within the extruded aluminium sections to reduce the incidence of heat conduction through the metal frame, known as the cold bridge effect.

PVCu windows

These are manufactured from hollow extrusions of PVCu that are fusion welded at their mitred corners to create frames (see Figure 16.12).

Small to medium sized windows do not need reinforcing, but larger windows do need to be reinforced by inserting stiffening metal sections into the hollow frame sections.

Fixing windows

To ensure a good fit between the window frame and the wall it is preferable to build windows in as the construction progresses. The frames may be secured to the wall by means of galvanised steel frame cramps (see Figure 16.13) or the frames may be screwed to wooden or plastic plugs that have been inserted into the mortar joints of the wall.

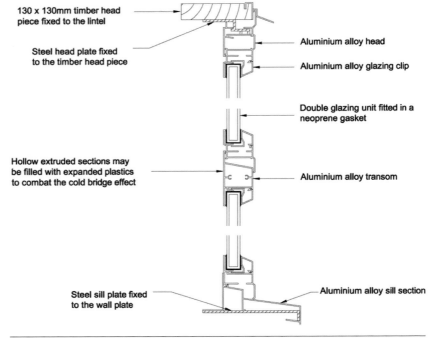

130 x 130mm timber head piece fixed to the lintel

Steel head plate fixed to the timber head piece

Aluminium alloy head

Aluminium alloy glazing clip

Double glazing unit fitted in a neoprene gasket

Hollow extruded sections may be filled with expanded plastics to combat the cold bridge effect

Aluminium alloy transom

Steel sill plate fixed to the wall plate

Aluminium alloy sill section

Figure 16.11 An aluminium window

The vertical DPC that is built into the wall reveals is fixed to the anti-capillary groove in the back of the window frame. The face of the gap between the frame and the wall is then sealed with mastic.

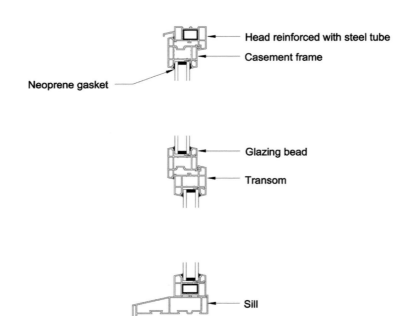

Head reinforced with steel tube

Casement frame

Neoprene gasket

Glazing bead

Transom

Sill

Figure 16.12 A PVCu window

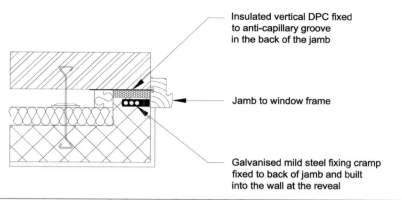

Insulated vertical DPC fixed
to anti-capillary groove
in the back of the jamb

Jamb to window frame

Galvanised mild steel fixing cramp
fixed to back of jamb and built
into the wall at the reveal

Figure 16.13 Fixing window frames with a frame cramp

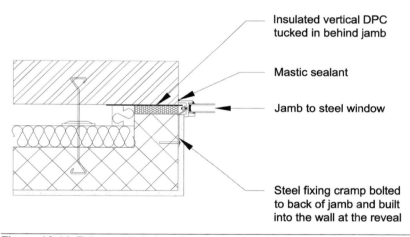

Insulated vertical DPC
tucked in behind jamb

Mastic sealant

Jamb to steel window

Steel fixing cramp bolted
to back of jamb and built
into the wall at the reveal

Figure 16.14 Fixing steel windows by means of steel fixing straps

Steel windows may be fixed directly to the wall reveals by means of steel fixing straps that are bolted to the back of the frame (see Figure 16.14). Alternatively they may be screwed into timber or plastic plugs inserted into the wall.

The steel window may also be fixed to a timber sub-frame, which has been built into the wall reveals (see Figure 16.15).

Aluminium and PVCu windows are normally fixed into prepared openings, formed around a temporary timber template. The windows may either be fixed into plugs inserted into the wall reveals or aluminium frames may be fitted into a timber sub-frame in a similar manner to that of steel windows.

Glazing

Glass is produced by fusing together silica, lime, magnesia, alumina and iron oxide at approximately 1500^0–1550^0C. It is then formed into sheets

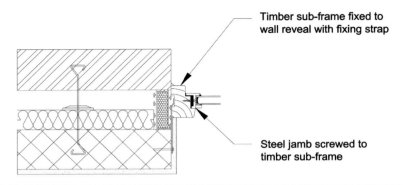

Timber sub-frame fixed to wall reveal with fixing strap

Steel jamb screwed to timber sub-frame

Figure 16.15 Fixing steel windows to a timber sub-frame

by a process of drawing, floating or rolling. The most common types of glass used for building are:

- Flat drawn – the cheapest method of production. A considerable width of molten glass is drawn up a tower from the furnace between rollers. Once it has cooled it is cut into sheets of suitable sizes.
- Float – molten glass is passed over a bed of molten tin. This produces a sheet of more uniform thickness than that produced by drawing.
- Polished plate – this is produced as a ribbon through horizontal rollers and then passes through an *annealing* furnace to strengthen it.
- Obscured – one surface of the glass is patterned as it passes between rollers. The pattern obscures direct vision through the glass but still provides adequate daylight admission.
- Georgian wired – the addition of wire mesh reinforcement to glass at the rolling stage helps to hold the glass together should it become shattered. Georgian wired glass has some fire resistance, but it is not any stronger than normal glass and is not impact resistant.

Apart from these main forms of glass others are also available:

- Toughened – if the glass is super cooled then any stresses set up by the cooling process may counterbalance any later stresses on the glass that may be set up by impact or rapid changes in temperature. This super cooled glass is often referred to as *armour plate glass*. Toughened glass cannot be cut, as this would destroy the inbuilt stresses within the glass. It must therefore be ordered to the correct size.
- Laminated – two or more layers of glass are bonded together with plastic or inorganic layers. These layers absorb energy during impact. Fracture of the sheet does not cause the glass to separate from the interlayer. It is heavier and more expensive than toughened glass but can be cut to size. It is available as safety glass, anti-bandit glass and bullet resistant glass. Laminated glass is efficient as a sound insulator since the interlayer gives reduction of sound transmission at different frequencies to that of the glass. The glass is normally transparent but becomes opaque at high temperatures, thus reducing transmission of radiant heat.

Self-assessment question 16.5

How can the change of laminated glass at high temperatures from transparent to opaque be of particular use in buildings?

- Coated – the deposition of various coatings on the surface of glass can help to alter its performance characteristics. Low E glass has a metal or metal oxide coating on the outer face of the inner pane in a multiple glazed unit having a low heat emissivity. Short wave radiation from the sun can enter the building but long wave radiation, from the heat inside the building, is reflected back into the room. There are two types of coatings that can be used; hard coatings are applied during the glass manufacturing process whilst soft coatings are applied after manufacture. Although the soft coatings have lower emissivity values, they tend to degrade when exposed to air and moisture and are easily damaged. Tinted glasses provide glare control. Coloured coatings can create various visual effects.

Glass may be held in position within its rebates by means of glazing beads, putty or neoprene gaskets. All openings to receive glass should have a minimum rebate of 5mm and the glass should be cut 3mm shorter in length and breadth than the actual size of the opening.

Self-assessment question 16.6

Why should glass be cut 3mm shorter in length and breadth than the actual size of the opening?

Glazing with putty

This is normally used on timber and steel windows. The rebate to the window frame is puttied. The glass sheet is pressed into position and fixed with *glazier's sprigs*. The glass sheet is then front puttied. The putty is normally stopped 2mm from the sight line of the rebate to allow painting over to seal its edge. The putty hardens by the absorption of its linseed oil into the timber window frame. In steel windows no such absorption takes place so the putty must incorporate a hardening agent.

Glazing with beads

This is generally performed on internal glazing but can also be used on steel windows. The glass is placed into the window rebate against putty, chamois leather or velvet. The beads are fixed into the window frame by means of pins or screws with cup washers. Steel windows use steel beads secured with self-tapping screws.

Glazing with neoprene gaskets

These are mainly used on aluminium and plastic windows. The gasket contains the seal for the glazing and is accommodated within a rebate within the frame of the window. It provides good weather resistance and is easy to remove, enabling re-glazing to be carried out swiftly.

Window furniture

There are two main items of window furniture:

- the casement fastener
- the casement stay.

The casement fastener

This is a security device to fix the casement into its closed position. It is normally operated by a hand-operated blade fixed to the edge of the opening casement and locating over a projecting wedge on a plate fixed to the mullion (see Figure 16.16). Many modern casement fasteners are supplied with locking devices to provide extra security to the window.

The casement stay

This is a fixing device designed to hold the casement in any number of desired opening positions. It comprises an arm with a series of locating holes and is fixed to the bottom edge of the opening casement. A locating pin on a plate is fixed to the cill member (see Figure 16.17). Many modern windows now incorporate friction stays that do the same job as the casement stay.

Window furniture to timber windows is generally fixed after the windows have been installed. Steel, aluminium and PVCu windows are normally supplied with the window furniture already fixed.

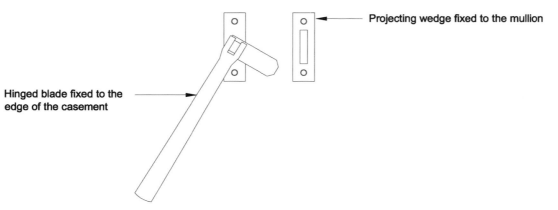

Projecting wedge fixed to the mullion

Hinged blade fixed to the edge of the casement

Figure 16.16 A casement fastener

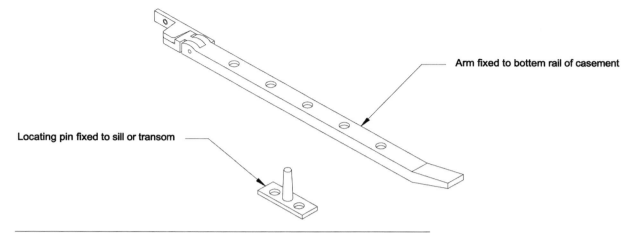

Arm fixed to bottom rail of casement

Locating pin fixed to sill or transom

Figure 16.17 A casement stay

 Visit the companion website to test your understanding of Chapter 16 with a multiple choice questionnaire, and to see the following features illustrated with full-colour photos:

- **16.1** A PVCu window
- **16.2** Fixing of a PVCu window frame to the inner leaf of a cavity wall
- **16.3** Opening casements to a PVCu window
- **16.4** Timber windows
- **16.5** Top hung metal window with metal sill section

CHAPTER 17

Doors

Introduction

Doors, like windows, may be constructed from timber, steel, aluminium and PVCu. The material used for the construction of the door will be influenced by the functional requirements that need to be achieved, cost considerations and the location of the door within the building. External doors have different functional requirements to internal doors. Many external doors and the majority of internal doors will be constructed from timber.

Frames are generally supplied independently of the door and may be manufactured of similar materials to the door. However, PVCu doors are supplied in their associated frames and some internal timber doors are now being supplied hung in their associated frames as *doorsets*.

This chapter will consider the functional requirements of external and internal doors. It will then discuss the construction of timber panelled doors, timber flush doors, timber fire doors and timber matchboarded doors. Consideration will also be given to aluminium and PVCu external doors and their associated frames. Finally, door frames, door linings and door furniture for timber doors will be discussed.

Functional requirements

The functional requirements for doors will be dependent on the location of the door in the building. External doors will have different requirements to internal doors.

External doors need to provide:

- Weather resistance – as part of the external envelope of the building the door and its associated frame should not allow rain or air infiltration into the building. This can be difficult to achieve as the door needs to be able to be opened and closed easily, and therefore there needs to be a sufficient gap between the door and the frame to allow for this. This gap can be a potential point of entry for rainwater and draughts into the building and for heat to escape from the building. The design of the door and frame must therefore provide features that will overcome these potential problems as effectively as possible. The ability to retain the shape of the door when it is subjected to changes in climatic conditions is a major concern with timber doors. After all, they may be subjected to quite varying climatic conditions on the outside and inside faces at the same time. This shape stability tends to be better achieved on heavier timber doors than it is on their lighter counterparts. The addition of a metal foil vapour control layer behind the facings of flush doors helps to maintain shape stability.

Self-assessment question 17.1

Why should shape stability be better achieved on heavier timber doors than it is on lighter doors?

- Durability – the material that the door and its associated frame are manufactured from will affect their durability. In general, aluminium and PVCu doors will have a longer life span than timber doors but will also be more expensive to purchase initially. Because doors and frames are easier to replace than other more permanent parts of the building structure the expected life span of these components may be as short as 20–25 years before they are replaced. Timber doors, although inexpensive to purchase initially, do require regular protection to be applied to them in the form of paint, stain or sealant. This maintenance cost needs to be added to the initial cost of the door and frame before a true cost comparison can be made between timber doors and their aluminium and PVCu equivalents. This *life cycle costing* considers the total cost of a component over its entire life cycle and includes projected maintenance and replacement costs as well as the initial purchase and installation cost of the product. In this way a component that could have a low initial cost in comparison to an alternative component may have higher life cycle costs if it has high maintenance and replacement costs compared to the alternative material.

Self-assessment question 17.2

Which is likely to be better value: an anodised aluminium door in hardwood frame costing £600 to supply, £70 to fit and £3 per year to clean over a lifespan of 30 years or a hardwood panel door and frame costing £250 to supply, £40 to fit and £35 to decorate and re-decorate once every 4 years with a lifespan of 20 years?

- Strength and stability – because doors are the main entry and exit points into a house and to various rooms in the house they need to be strong enough to resist onslaught by intruders. Hardwood door frames are stronger than softwood frames since they have better resistance to splitting. The type of locks, bolts and latches that are fixed to the door also influences the security of doors. Strength of doors is measured by four criteria covered by *PAS 23-1:1999 General performance requirements for door assemblies. Single leaf, external door assemblies to dwellings*: resistance to torsion, resistance to closing against an obstruction, resistance to impact (heavy body) and resistance to impact (hard body). The strength of panelled doors is related to the sizes of the members used in their construction and the method of jointing these members. In flush doors, strength is closely related to the construction of the core of the door.

- Thermal insulation – as doors form a part of the external fabric of the building they will affect the amount of heat that may be lost from the building. Doors do not have the same level of thermal performance as the walls into which they are fitted. Doors that fit badly within their associated frames may suffer greater amounts of heat loss than those that fit well. Weatherstripping to door frames reduces draughts caused by air infiltration through the gaps between the door and its frame and also reduces heat loss through these gaps.
- Sound insulation – as with thermal insulation, doors will have poorer acoustic performance than the wall into which they are installed. In general heavier doors will provide better insulation against the transmission of airborne sound than will lighter doors.

Internal doors are only really required to provide:

- Fire resistance – in certain areas of the house, doors and the internal walls into which they are installed will act as a vertical division of space within the building, which can be used to reduce the spread of fire from one area of the building to another. This concept is known as *compartmentation* and is covered by the Building Regulations Approved Document B, Section 5. Doors that are installed in compartment walls and that open onto corridors that are designated as routes for the means of escape from the building in the event of a fire will need to be fire doors. These are doors that have materials incorporated into their construction that enable them to maintain their functional performance for a minimum set period during a fire. This period should enable people who are within a building when it catches fire to be able to exit from the building via designated escape routes to a place of safety. Fire doors are considered in more detail later in this chapter.
- Sound insulation – as internal doors provide a means of access and egress between rooms in a building they also need to be capable of preventing the transmission of sound from one room to another. Sound insulation of internal doors may also be required to maintain privacy. Doorsets, where the door is supplied with the associated frame, will generally have better sound performance than doors supplied separately from the frame, as the gap between the door and the frame is minimalised.

Because internal doors need to satisfy less functional requirements than external doors they can often be made of a much lighter construction and at less cost.

Timber doors

Timber doors may be classified into four types by their method of construction:

- panelled doors
- flush and moulded doors
- fire doors
- matchboarded doors.

Panelled doors

These may be of single or multiple panel construction, comprising a timber framework encasing panels of timber, plywood or glass to *BS 4787–1:1980 Internal and external wood doorsets, door leaves and frames. Specification for dimensional requirements* (see Figure 17.1).

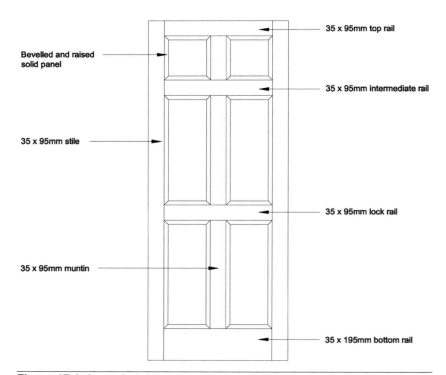

Bevelled and raised solid panel

35 x 95mm top rail

35 x 95mm intermediate rail

35 x 95mm stile

35 x 95mm lock rail

35 x 95mm muntin

35 x 195mm bottom rail

Figure 17.1 A panelled door

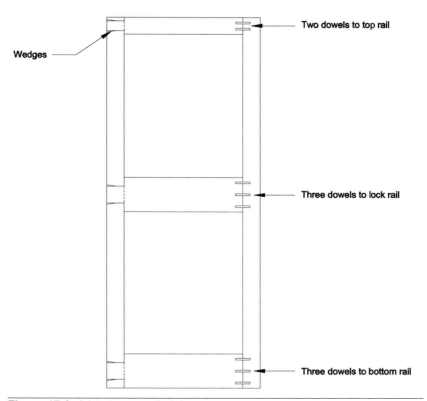

Wedges

Two dowels to top rail

Three dowels to lock rail

Three dowels to bottom rail

Figure 17.2 Joints to a panelled door

This type of door is mostly used externally, therefore if plywood panels are to be used the plywood should be class 3 to provide adequate weather resistance.

The joints used in the panel framing may be doweled or mortice and tenoned (see Figure 17.2). The dowel joint is more popular, due to its cheapness and ease of construction, which is particularly suited to factory production lines. However the mortice and tenon joint tends to be stronger. Where doweling is used there should be 3 dowels to the 195 x 45mm bottom and lock rails, 2 dowels to the 95 x 45mm top rail and 1 on any 95 x 45mm intermediate rails. Where mortice and tenon joints are used the middle and bottom rails should have double tenons.

The panels are framed into grooves within the framing stiles and rails and are covered by mouldings that are *scribed* at their corners.

Water that runs down the face of the door is thrown clear at the bottom by a 40 x 65mm *weatherboard*, which is attached to the face of the door, preferably by a tongue and grooved joint. The weatherboard has a sloping face and is grooved on its underside to prevent the underflow of water.

Where glazing is used, any glazing beads should be positioned on the inside of the door.

Self-assessment question 17.4

Why should glazing beads be positioned on the inside of external doors?

Flush and moulded doors

Traditionally flush doors have plain faces that are easy to clean and decorate, being free of the mouldings that collect dirt. Flush doors may be faced with plywood, medium density fibreboard (MDF) or hardboard and may be manufactured for internal or external use. For external use the facing to the doors must be of class 3 plywood and painted. For internal use the doors may be painted or supplied with a decorative veneer.

Moulded doors have their facing material made from MDF or hardboard that has been moulded under pressure to produce a decorative appearance similar to that of a panelled door. However, although the door is similar in appearance to a panelled door it does not have the same construction as a panelled door.

Self-assessment question 17.5

Why do moulded panel doors use MDF or hardboard as a facing and not plywood?

BS 4787 does not specify a method of construction for flush or moulded doors, thus manufacturers have freedom of designs for different price ranges. The method of construction for these doors tends to be of three distinct types, related to cost:

- hollow core
- semi-solid core
- solid core.

Hollow core

This comprises an outer frame of 125 x 29mm timber enclosing a core of polystyrene, plastic or expanded cellular paperboard construction, to which are attached MDF or hardboard facings (see Figure 17.3). The resulting door is light and inexpensive, but it can be difficult to achieve a flat surface to the facings due to the limited support offered by the core.

Semi-solid core

This comprises an outer frame of 125 x 29mm timber within which small 29 x 25mm timber intermediate rails are spaced to provide a minimum of

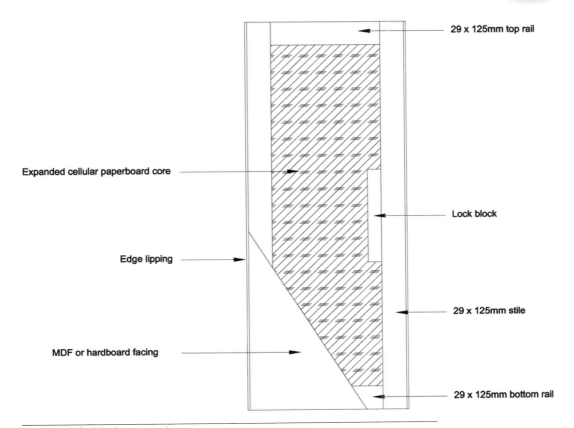

Figure 17.3 A hollow core door

50 per cent timber in its construction, over which are then fixed the facings of plywood, MDF or hardboard (see Figure 17.4). This door is still relatively light and inexpensive but the intermediate rails do provide more support for the facings so that a flat surface can be achieved more easily than with the hollow core door.

Solid core

This door comprises a solid core of either blockboard or laminboard onto which are attached the plywood facings (see Figure 17.5). Alternatively the core can consist of particleboard or flaxboard.

This door is of much heavier construction and consequently more expensive than the hollow core and semi-solid door types and has good sound insulation properties.

All flush doors and moulded panel doors may incorporate glazed panels, secured by fixing beads.

The facings to these doors tend to be vulnerable at their edges, so a thin strip of timber is fixed at the edges to provide a *lipping*. This also provides a tolerance for planing the doors to fit the frame or door lining when the doors are hung into position.

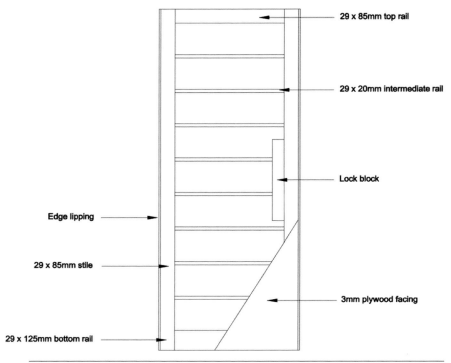

29 x 85mm top rail

29 x 20mm intermediate rail

Lock block

Edge lipping

29 x 85mm stile

3mm plywood facing

29 x 125mm bottom rail

Figure 17.4 A semi-solid core door

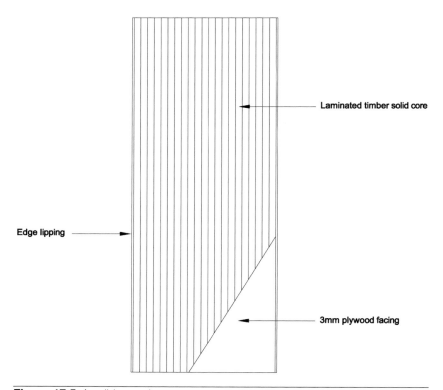

Laminated timber solid core

Edge lipping

3mm plywood facing

Figure 17.5 A solid core door

Fire doors

The main requirement for a fire door in a house is where access is provided between the house and an attached garage. The Building Regulations Approved Document B, Section 5.5 requires that such a door should have 30 minutes fire resistance.

BS 5588-1:1990 Fire precautions in the design, construction and use of buildings. Code of practice for residential buildings specifies codes for the various categories of fire door. This code uses the prefix FD for all fire doors, followed by the suffix 20, 30, 60 or 120 to denote the period of fire resistance provided. These periods of fire resistance relate to *BS 476-22:1987 Fire tests on building materials and structures. Methods for determination of the fire resistance of non-loadbearing elements of construction*. This publication rates the fire resistance of doors by their performance in the *integrity* test. Integrity is related to the ability of the door to resist the passage of flame for the designated period of fire resistance for the category of the door. Because the performance of the fire door needs to take into account not only the door leaf but also the associated door frame and door furniture, the integrity test is performed on the complete assembly of door, frame and furniture.

Self-assessment question 17.6

Why should the fire tests laid down in Parts 22 and 31 of *BS 476* require door leaves to be tested with their frames and furniture?

It is still not common practice however for the door manufacturer to supply fire doors complete with their frame and furniture. It is therefore possible for the performance of a fire door to be affected by the choice of frame and furniture by the builder and their installation on site.

Self-assessment question 17.7

Why should the performance of a fire door assembly be materially affected by the choice of frame and furniture by the builder and their installation on site?

For this reason certification schemes have been developed by fire door manufacturers to ensure that all components of the fire door assembly, including the installation, can be carefully monitored to ensure compatibility with the requirements laid down by the British Standards. This also would enable fire door manufacturers to supply the door leaf alone and for others to supply the door frame and furniture if desired.

In addition, *BS 5588* also provides for a further category of fire doors where restricted smoke leakage at ambient temperatures is required. These doors are identified by the further suffix 's' following the period of fire resistance designated for the door. Tests for smoke penetration for these doors is covered by *BS 476-31:1983 Fire tests on building materials and structures. Methods for measuring smoke penetration through doorsets and shutter assemblies. Method of measurement under ambient temperature conditions.* These tests take account of temperatures at three different stages of a fire. The Building Regulations consider the performance evaluated under *BS 476-31.1 Measurement under ambient temperature conditions*.

There are also European Standards for fire resistance; *BS EN 1634-1:2008 Fire resistance and smoke control tests for door, shutter and openable window assemblies and elements of building hardware. Fire resistance tests for doors, shutters and openable windows,* and *BS EN 1634-3:2004 Fire resistance and smoke control tests for door and shutter assemblies, openable windows and elements of building hardware. Smoke control test door and shutter assemblies*. These regulations go one step further than the requirements laid down in *BS 476*, in that they not only classify a fire door by its integrity rating, but also by its insulation rating. This is the time, in minutes, for the temperature of the face of the door facing away from the fire to equal the temperature of the face of the door facing the fire.

The most important criterion for designating the period of fire resistance related to the passage of flame or smoke through the fire door assembly is the size of the gap between the door leaf and its associated frame. This gap needs to be wide enough to enable the door to be opened or closed easily during normal use, but not too wide as to allow the passage of flame or smoke in a fire. It is not possible for a timber fire door of normal thickness and with normal gaps between the door leaf and the frame to provide integrity of more than 20 minutes. This problem has been alleviated considerably by the use of *intumescent strips* fitted into grooves within the edge of the door leaf. This consists of a chemical that remains inactive at temperatures below those normally experienced in a fire, but will expand on being heated to a high temperature, such as that sustained in a fire. When the strip expands it effectively seals the gap between the door leaf and the frame thus reducing the ability for flames, smoke and gasses to penetrate through the gap.

There are three main types of material used for intumescent strips. Ammonium phosphate becomes active at 180^{0}C, hydrated sodium silicate becomes active at 120^{0}C and intercollated graphite becomes active at 200^{0}C. The first two are hygroscopic in nature and therefore have to be encased in PVC to prevent them from degradation through contact with moisture. Some fire door manufacturers have incorporated the intumescent strip within the door construction at the edge, just beneath the lipping in order to protect it. In these doors the glue joint fixing the lipping to the door leaf edge must be designed to soften when heated to enable the lipping to be forced off by the swelling of the intumescent strip so that it can seal the gap between the door leaf and the frame.

The Building Regulations Approved Document B, Section 5.5 stipulates that a fire door between the house and an attached garage should be of FD30s designation (E30Sa under the European regulations). As with flush doors, there is no British Standard Specification for the construction of fire doors, thus door manufacturers are free to design the door leaf as they wish, provided it meets the requirements of the fire tests laid down in parts 22 and 31 of *BS 476* or *BS EN 1634* parts 1 and 3. Most fire doors comprise a thick central core, either a slab of particleboard or flaxboard or a slab of these materials supported by framing to provide increased stability. In order to improve the fire resistance of a flush door or moulded panel door construction, fire resisting materials such as plasterboard are often placed over the core construction and covered by the facings (see Figure 17.6). The doors are a minimum of 45mm in thickness for FD30 designation and 54mm thick for FD60 designation.

Frames to fire doors must also be designed to meet the requirements of the fire tests laid down in *BS 476*. These will be considered later in this chapter.

Matchboarded doors

These doors are constructed to the requirements of *BS 459:1988 Specification for matchboarded wooden door leaves for external use*, and are available in three specific types:

- ledged and braced
- framed and ledged
- framed, ledged and braced.

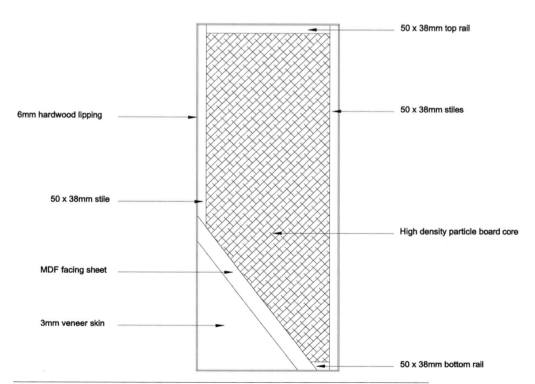

Figure 17.6 A typical FD30 fire door

Ledged and braced

The face of this door is constructed from 16mm thick tongued and grooved boarding, called *battens*, with edge chamfers to form a 'V' joint feature between consecutive boards. The battens are clamped together to form a panel and are attached by nailing to 95 x 22mm horizontal timbers, called *ledges*, positioned in the middle of the door panel and also 152mm from its top and bottom edges.

To resist the tendency for the door panel to drop out of square, 95 x 22mm diagonal braces are fitted between the ledges. The braces are fixed to the ledges by means of mortice and tenon joints (see Figure 17.7).

To further resist the tendency for the door to drop out of square, three hinges, rather than two, are fixed to the ledges on the hanging side of the door. *Tee hinges* (see Figure 17.8) are often used on this type of door.

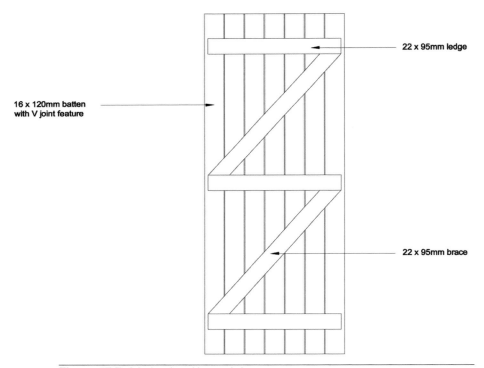

16 x 120mm batten with V joint feature

22 x 95mm ledge

22 x 95mm brace

Figure 17.7 A ledged and braced door

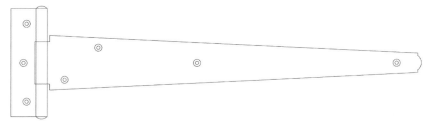

Figure 17.8 A tee hinge

Framed and ledged

Added strength and rigidity can be provided to the door by enclosing the panel in a mortice and tenoned frame comprising 95 x 44mm top rail and stiles, 146 x 27mm middle and bottom rails. The vertical battens cover the middle rail and bottom rails and are rebated into the top rail and stiles.

Framed, ledged and braced

This is similar to the framed and ledged door, but with the addition of 95 x 27mm braces to provide further rigidity, particularly for wide doors. The joints to the framing members are mortice and tenoned or dowelled.

Matchboarded doors are now very rarely used for internal or external doors on houses but are still specified for outbuildings and garages. The framed ledged and braced door is particularly suitable for use as garage doors as their frame construction makes them stronger than the other types of matchboarded doors for particularly wide openings.

Aluminium doors

These are constructed from extruded aluminium alloy sections in a similar manner to aluminium windows. There is no separate British Standard Specification for aluminium doors but *BS 4873:2009 Aluminium alloy windows and doorsets. Specification* relates to door construction too. The doors normally comprise top, middle and bottom rails and stiles of extruded aluminium alloy sections enclosing glazed panels. The finishes to the doors are similar to those available for aluminium windows.

The doors are normally fitted into hardwood frames supplied separately.

PVCu doors

Similar to aluminium doors, PVCu doors are constructed from rigid PVCu hollow extrusions in a similar manner to PVCu windows to *BS 7412:2007 Specification for windows and doorsets made from unplasticized polyvinylchloride (PVC-U) extruded hollow profiles*. Large doors may require these extrusions to be stiffened with steel sections inserted into the hollow cores. The doors comprise top, intermediate, middle and bottom rails and stiles of extruded PVCu sections enclosing PVCu panels and glazing according to design. They are generally supplied in white but other colours are available, including timber effect.

The doors are fitted into PVCu frames supplied with the doors.

Door frames

A frame is generally used for external doors, heavy doors and fire doors. It consists of four members, a 85 x 57mm head, a pair of 85 x 57mm

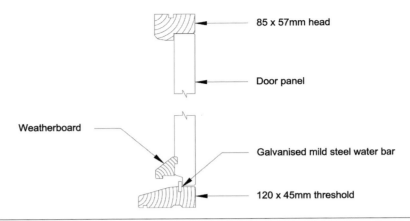

85 x 57mm head

Door panel

Weatherboard

Galvanised mild steel water bar

120 x 45mm threshold

Figure 17.9 A door frame

vertical jambs and a 120 x 45mm *threshold* or sill, joined together by doweled, mortice and tenon joints or *combed joints* (see Figure 17.9).

All fire doors must be fixed into timber frames with a minimum 12mm rebate and a maximum gap of 3 mm between the door and the frame. Frames for FD30 fire doors may accommodate a *planted stop*, which must be a minimum of 12mm in thickness. The density of timber used for the frame should be a minimum of 480kg/m³ for FD30 fire doors and 650kg/m³ for FD60 fire doors.

To prevent the ingress of water beneath external doors, a galvanised mild steel water bar can be placed within the threshold of the frame and a rebate is cut in the bottom rail of the door to ensure a tight fit against the water bar. The top surface of the threshold to the exterior of the water bar is sloped to allow water to be conducted away from the opening. The underside of the threshold is throated to prevent water from running underneath it (see Figure 17.9).

Recent legislation concerning access into buildings for disabled people has caused a re-think on the threshold detail for buildings where access for wheelchair users is required. This entails separating the threshold from the rest of the timber frame and installing a stainless steel threshold section to the external wall below the door opening. A complementary water bar with draughtproofing seals can be fixed to the bottom of the door to prevent the ingress of rainwater and draughts.

The frames may be built in to the wall reveals, as the wall is being constructed, by means of galvanised mild steel frame cramps, similar to that used for fixing timber windows. Alternatively the frame may be secured by screwing into plugs inserted in the reveals.

Door linings

A door lining is normally used for internal doors. It consists of 25–32mm thick timber boarding with joints housed together and a 35 x 12mm planted stop fixed to house the door in the closed position (see Figure 17.10).

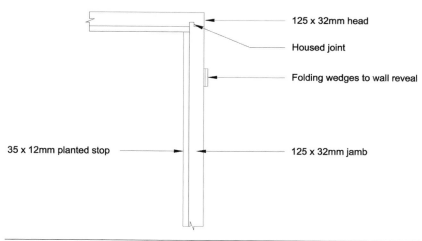

Figure 17.10 A door lining

Door linings are never built in, but are fixed into a prepared opening by fixing them directly to the reveals. Timber packing pieces are used to straighten and plumb up the jambs prior to the fixing of the door. The gap between the door lining and the reveals is covered by an *architrave*.

Doorsets

Traditionally door leaves were supplied independently of their associated frames or linings, which were manufactured separately. When doors were fitted into their frames or linings, there was generally a need to plane the edges of the door in order to ensure that they fitted properly. In recent years door manufacturers have been supplying doors fitted within their associated frames. These doorsets have the door fixed to the frame using *rising butt hinges* (see Figure 17.11), so that the door can be taken out of the frame prior to the frame being fitted into the wall and then lifted back into position when the finishes to the wall have been completed.

Because the door is delivered to site fitted in its frame there is no need for subsequent planing of the door to fit it into the frame when the door is finally installed.

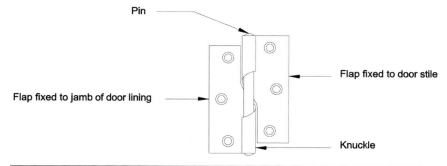

Figure 17.11 A rising butt hinge

Door furniture

There are a wide variety of items that can be fixed to the door for operational, security or durability purposes. The main items of door furniture used in domestic properties are:

- hinges
- latches
- locks
- bolts
- letter plates
- finger and kicker plates
- self-closing devices.

Hinges

These are normally supplied in pairs, although heavier doors may need three hinges. The main types of hinge are:

- *Butt hinge* – this is the simplest type of door hinge, comprising two flaps held together by a pin (see Figure 17.12).
- *Rising butt hinge* – this is similar to the butt hinge but the flaps are spiral in shape where they are attached to the pin (see Figure 17.11). This enables the door to lift when opened and to self-close.
- *Tee hinge* – this hinge has an elongated flap that allows the weight of the door to be distributed over a wider area of the hinge (see Figure 17.8).

Latches

These are mechanisms to secure the door in the closed position. They are generally operated by a lever handle to *BS EN 1906:2010 Building hardware. Lever handles and knob furniture. Requirements and test methods*, operating a central spindle that pulls the latch bolt back within

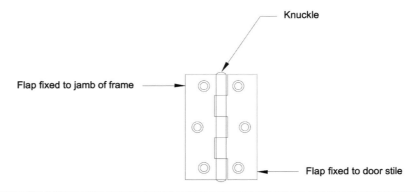

Knuckle

Flap fixed to jamb of frame

Flap fixed to door stile

Figure 17.12 A butt hinge

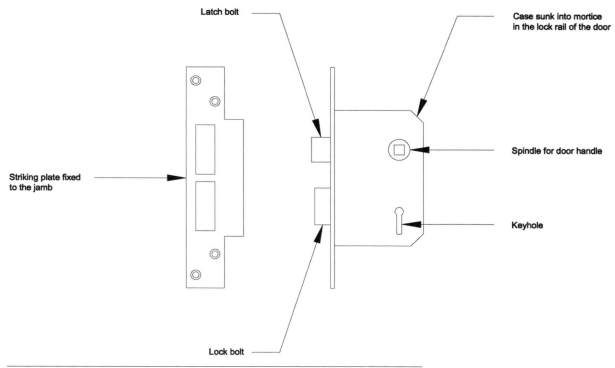

Latch bolt

Case sunk into mortice
in the lock rail of the door

Striking plate fixed
to the jamb

Spindle for door handle

Keyhole

Lock bolt

Figure 17.13 A door latch and lock

its case by the aid of a spring. When the latch bolt is allowed to extend
back into the closed position it is located into a mortice cut into the jamb
of the door frame or lining. The mortice is covered by a keep with a
striking plate that guides the latch bolt into the mortice (see Figure 17.13).

On cellular core and semi-solid core flush and moulded panel
doors a block of timber, called a lock block, is provided within the door
construction to enable the latch mechanism and handles to be attached
to the door.

Self-assessment question 17.8

Why are lock blocks not used on solid flush doors?

Locks

These are similar to latches but are generally operated by a key rather
than by a handle. Most external doors will have a combined latch and
lock mechanism (see Figure 17.13). Five lever locks are more secure than
three lever locks.

Cylindrical night locks to *BS EN 1303:2005 Building hardware.
Cylinders for locks. Requirements and test methods* are fitted to the stile
of the door. Their keep is attached to the jamb of the frame. These locks

have a safety mechanism that can be activated from the inside to prevent the lock from being opened from the outside.

Bolts

These are security devices consisting of a barrel sliding within a plate fixed to the inside of the door and a keep, into which the barrel is slid, attached to the jamb of the frame. They are covered *by BS EN 12051:2000 Building hardware. Door and window bolts. Requirements and test methods*.

Letter plates

These are hinged covers fixed over an opening of recommended size 200 x 45mm, bolted through the door and secured by nuts on the inside face of the door.

Finger and kicker plates

These protect the face of the door from damage at vulnerable points and are screwed to the face of the door at the desired position. Again with flush and moulded panel doors, extra blocking can be incorporated into the door construction to enable these plates to be fixed to the door.

Self-closing devices

Fire doors must be kept in the closed position when not in use. It is therefore necessary to fit a self-closing device to the door to ensure that the door is not able to be left in the open position when not being used. Modern self-closing devices are quite unobtrusive and consist of a spring loaded chain mechanism that can be fitted through the stile into the core construction of the fire door and attached to the jamb of the frame.

 Visit the companion website to test your understanding of Chapter 17 with a multiple choice questionnaire, and to see the following features illustrated with full-colour photos:

- **17.1** GRP external moulded door
- **17.2** Timber moulded external door

Stairs

Introduction

The primary function of stairs is to provide a means of access from one floor level to another, but they must also provide a means of escape in case of fire.

Stairs in houses are normally constructed from timber and a number of designs are used, dependent on the layout of the house and the space available for the staircase.

This chapter will first consider the functional requirements of a staircase and then consider the factors affecting staircase design. Finally the construction of a straight flight staircase for a typical house will be discussed.

The functional requirements of stairs

Stairs should satisfy the following functional requirements:

* strength and stability
* fire resistance
* sound insulation.

Strength and stability

Stairs, like floors, must carry the weight of the people using them, plus the weight of any furniture and equipment that may be transported over them. However, no live loads are imposed, as furniture and equipment will not be located on the stairs, as they will be on floors.

As with floors the loads on stairs can be related to the use classification of the building.

Fire resistance

The construction of the stairs must allow them to maintain their functional use during the early stages of a fire, to enable occupants of the building to escape to a place of safety and, if possible, to provide fire fighting services with access to the upper storeys of the building.

The fire resistance of the stairs will be dependent upon the use classification of the building. In domestic construction, timber staircases will provide half-hour fire resistance, which will be adequate.

Sound insulation

Sound insulation will be necessary on the stairs in order to reduce sound transmission from one storey to another. The main form of sound transmission on stairs is structure borne or impact sound.

The most effective form of insulation against this is to provide sound absorbent finishes to the treads of the stairs. Unfortunately the open space that staircases occupy between storeys in a building (called the *stairwell*) carries airborne sound particularly well. There is very little that can be done to reduce this problem apart from placing sound absorbent finishes on the surrounding walls. These finishes will need to have a low surface spread of flame rating.

Staircase design

The design of a staircase is dictated by the space into which it is to be fitted. The total height of a stairway is dictated by the height between the floors it is connecting, and so the total horizontal distance that the stairway occupies (often referred to as the total *going*) is practically the only dimension that can be set by the designer. However, the space available to construct the staircase may prevent the stairway from being designed in a single *flight* of steps. Thus the designer may need to turn the stairway through 90° by the provision of a *quarter space landing* or 180° by the provision of a *half space landing* or 270° or even 360° by the combination of quarter space or half space landings. Indeed, it is also possible for the designer to use *tapered treads* to turn the stairway through an angle and still maintain a progression of steps (see Figure 18.1).

The designer is further constrained by the requirements of the Building Regulations Approved Document K1. *BS 585-1:1989 Wood stairs. Specification for stairs with closed risers for domestic use, including straight and winder flights and quarter or half landings*, also provides valuable advice on the design of stairs for domestic use.

The Building Regulations requirements

These are based on ergonomic and anthropometric data to ensure that the staircase is designed in such a way that it is easy and safe to use. Ergonomics is essentially the study of the relationship between workers and their environment, especially the equipment they use. Anthropometrics is the comparative study of the sizes and proportions of the human body.

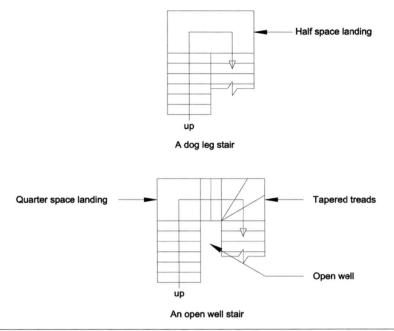

Figure 18.1 Landings and tapered treads

All buildings are classified by the Building Regulations into purpose groups according to the use of the building. Residential buildings are classified within the Regulations as requiring 'private' stairs. Only the requirements for private stairs will be considered in this chapter.

A step generally consists of two components, the vertical component or *riser* and the horizontal component or *tread* (see Figure 18.3). Some staircases do not contain risers; these are known as *open riser stairs* (see Figure 18.2).

Irrespective of whether the staircase contains risers or not, the vertical distance between two consecutive treads is referred to as the *rise*. This should be a maximum of 220mm (Building Regulations Approved Document K1, Section 1 Table 1*)*. In open riser stairs where the staircase is likely to be used by children under 5 years old the gap between two consecutive treads should not be greater than 100mm (Building Regulations Approved Document K1, Section 1.9*)*. This is so that a small child cannot crawl through the gap between the treads and fall from the back of the stairs (see Figure 18.2). To enable the rise to be greater than 100mm on these stairs it would be necessary to add a timber upstand to the top of the lower tread at its back or the underside of the upper tread to reduce the open space between them to 100mm (see Figure 18.2).

On most stairs there is an overlap between consecutive treads. This overlap is called the *nosing* (see Figure 18.5). On open riser stairs the length of this nosing must be at least 16mm (Building Regulations Approved Document K1, Section 1.8). Where nosings are used, the useable part of the tread will become the distance between the nosings on consecutive treads. This distance is called the *going* (see Figure 18.3).

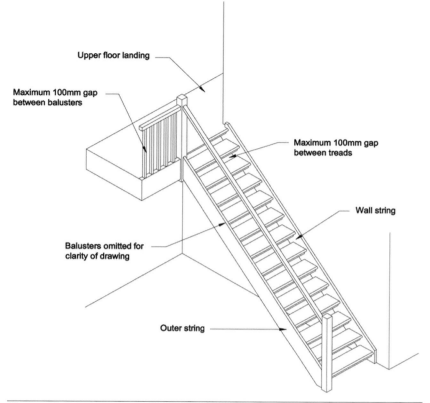

Figure 18.2 An open riser stair

This should be a minimum of 220mm (Building Regulations Approved Document K1, Section 1 Table 1).

Further stipulations about the dimensions of rises and goings are provided in Building Regulations Approved Document K1, Section 1.3. In a private stair the rise may be between 155mm–220mm with a going between 245mm–260mm. Alternatively the rise may be between 165mm–200mm with a going between 223mm–300mm.

The relationship between the rise and the going is further defined by the Building Regulations Approved Document K1, Section 1.5 as:

Twice the rise plus the going (2R + G) must be between 550–700mm where R is the rise and G the going of the stairway.

Furthermore, all goings should be of the same depth and all rises of the same height throughout the stairway.

Self-assessment question 18.1

Why should all goings be of the same depth and all rises be of the same height throughout the stairway?

On stairways where tapered treads are used (see Figure 18.1) the going should be measured in the middle of the tread where the width of the flight is less than 1m, and 270mm in from each side where the width of the flight is greater than 1m (Building Regulations Approved Document K1, Section 1.18). Furthermore, where a stair consists of straight and tapered treads, the going of the tapered treads should not be less than the going of the straight flight (Building Regulations Approved Document K1, Section 1.20).

The angle of inclination of the stairs is known as the *pitch* (see Figure 18.3). This can be measured as the angle between the floor at the foot of the stairs and a notional line connecting the nosings of all the treads (known as the *pitch line*). This should be a maximum of 42^0 (Building Regulations Approved Document K1, Section 1.4).

Self-assessment question 18.2

A stairway has been designed with a rise of 220 mm and a going of 220 mm. Does it comply with the requirements of the Building Regulations Approved Document K1?

This provides an envelope of acceptable dimensions for the rise and going that will satisfy the requirements of the Building Regulations Approved Document K1, Sections 1.4 and 1.5.

The *headroom* is the vertical distance between the pitch line of the stairway and the underside of the floor, landing or stairway above (see Figure 18.3). This needs to be a minimum of 2m (Building Regulations Approved Document K1, Section 1.10).

Self-assessment question 18.3

Why should the headroom be a minimum height of 2m?

However, where the staircase gives access to a loft conversion there may not be enough space to achieve the 2m minimum headroom height. In these cases the headroom is satisfactory if it is 1.9m at the centre of the stair width and at least 1.8m at the side of the stair.

The *flight* is a series of continuous steps that are not interrupted by a landing (see Figure 18.3). There is no minimum width of flight required by the Building Regulations but they should be wide enough to allow a satisfactory means of escape in case of fire and may also need to meet the requirements of Building Regulations Approved Document M where access for disabled people is required. There should not be more than 36 risers in consecutive flights without a change in direction between flights of at least 30^0 (Building Regulations Approved Document K1, Section 1.14).

Self-assessment question 18.4

Why should the number of steps in consecutive flights without a change in direction between flights of at least 30⁰ be restricted to 36?

The *landing* is the horizontal platform between flights (see Figure 18.1). The width and length of each landing should be at least as great as the smallest width of the flight it is connecting (Building Regulations Approved Document K1, Section 1.15) and should be positioned at the top and bottom of every flight. The landing may include part of the upper floor. Landings should be clear of any permanent obstruction (such as furniture). A door may be situated adjacent to a landing at the bottom of a flight of stairs, but only if it leaves a clear space of 400mm across the length of the landing when opened (Building Regulations Approved Document K1, Section 1.16).

The *handrail* is a protecting and safety member that is positioned at a convenient height usually on the outside edge of the staircase (see Figure 18.3). It assists people in climbing and descending the stairs but also prevents people from falling over the edge of the stairs. It should be at a height of between 900–1000mm above the pitch line and should be located on at least one side of the stair where the width of the stair is less than 1m and on both sides where the width of the stair exceeds 1m (Building Regulations Approved Document K1, Section 1.27).

Self-assessment question 18.5

Why should handrails need to be located on both sides of the stairs when the width of the stairs exceeds 1m?

Between the handrail and the steps there should be guarding on the outside of the stairs. This is normally provided by *balusters*, which can be fixed vertically between the handrail and the outer string. They are housed into the underside of the handrail at their head and the upper surface of the string at their foot (see Figure 18.4). The space between the balusters should not be greater than 100mm to prevent a child from becoming trapped between the balusters (Building Regulations Approved Document K1, Section 1.29). Guarding will also be required along the open side of any landing and the height of this guarding should be a minimum of 900mm (Building Regulations Approved Document K1, Section 3.2).

Construction of straight flight timber stairs

The 38mm thick treads and 20mm thick risers of the stair span between supports on either side of the stair, called *strings,* and are housed into the 38mm thick string member (see Figure 18.4). The inner string is normally fixed to a loadbearing wall, whilst the outer string spans between two

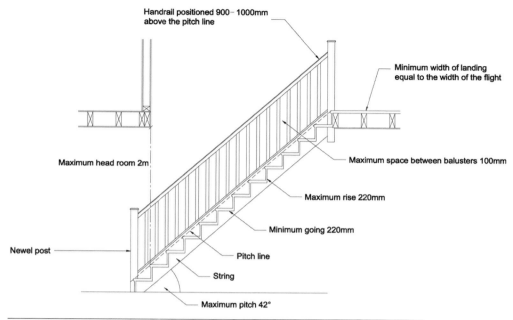

Figure 18.3 Dimensions of a stair way

upright 100 x 100mm *newel posts* at the head and foot of the stairs (see Figure 18.3).

The bottom newel post is normally fixed to the ground floor, whilst the top newel post is fixed to the stairwell trimming joist, which is part of the upper floor construction (see Figure 18.4). The outer string is tenoned to fit into prepared mortices in these newel posts (see Figure 18.4). The

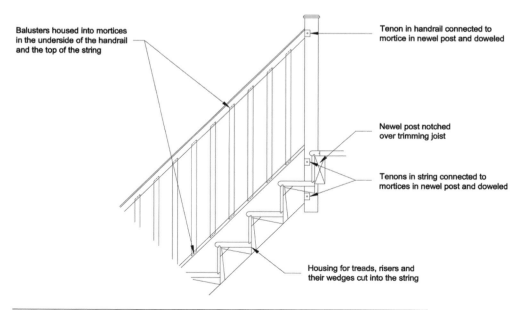

Figure 18.4 Joining of the string to the upper newel post

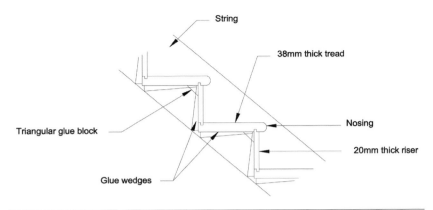

Figure 18.5 Joining of the treads and risers to the string

newel posts also support the outer handrail, which is also secured to them by a mortice and tenon joint.

The joint between the treads and risers is normally effected by providing a grooved joint in the bottom surface of the upper tread and the top surface of the lower tread, in which a tongue on each end of the riser will fit. Small triangular blocks are also glued to the underside of the treads and risers to further strengthen this intersection (see Figure 18.5).

The joint between the treads, the risers and the strings is strengthened by wedging and gluing them into position in their prepared housings (see Figure 18.5).

Rough brackets on the centre line of the stairs are glued to the underside of the treads and fixed to side bearers to provide further support to the staircase (see Figure 18.6).

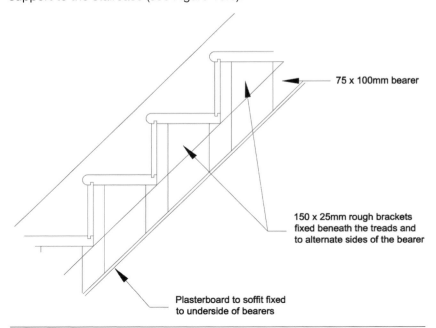

Figure 18.6 Rough brackets and bearers

Self-assessment question 18.6

Why should the rough brackets and bearers need to provide further support to the staircase?

At the landing the top riser is packed to the trimming joist. A small section of tread is located on top of the trimming joist and this picks up the tongue of the last riser on its underside and the tongue of the first floor board of the upper landing at its back (see Figure 18.4).

The wall (inner) string is taken over the trimming joist before being terminated at its junction with the *skirting board* on the upper landing.

To completely enclose the stairs, a soffit of plasterboard is fixed to the bearers on the underside of the stairs.

 Visit the companion website to test your understanding of Chapter 18 with a multiple choice questionnaire, and to see the following features illustrated with full-colour photos:

- **18.1** A newel post at the foot of a stair
- **18.2** A newel post fixed to a loadbearing wall and providing support to the outer string in a wider stair
- **18.3** A soffit to a timber staircase
- **18.4** A straight flight stair supported by an internal loadbearing wall
- **18.5** A straight flight stair
- **18.6** Sound insulation to the underside of the stairs
- **18.7** Tapered treads to a winder stair
- **18.8** The connection of the inner string to the internal loadbearing wall
- **18.9** The outer string to a stair supporting the ends of the treads and risers
- **18.10** The underside of a staircase showing the construction
- **18.11** Triangular blocks used to strengthen the connection between the treads and risers
- **18.12** Winder stairs

CHAPTER 19

Plasters and plasterboards

Introduction

Plasters and plasterboards are normally used to provide finishes to internal walls and ceilings. As such they provide additional attributes to the wall or ceiling, such as:

- the provision of a suitable surface for decoration
- masking unevenness in the background
- improving the sound insulation of the element
- improving the thermal insulation of the element
- improving the fire resistance of the element
- reducing condensation by improving the ability of the surface to absorb moisture.

There are many types of plasters and plasterboards and these have been developed to provide special characteristics that enable them to be used in specific circumstances. This chapter will consider the composition and types of plasters, the use of expanded metal lathing for plastering, the application of plasters and finally the composition, types and uses of plasterboard.

The composition and types of plasters

Modern plasters are manufactured from *gypsum*, a naturally occurring mineral that consists of a crystalline combination of calcium sulphate and water ($Ca\,So_4 \times 2H_2O$).

The gypsum is crushed and then heated to remove as much of the water of crystallisation as possible. The amount of water removed by this process defines the class of plaster, as laid down in *BS EN 13279-1:2008 Gypsum binders and gypsum plasters. Definitions and requirements*. If powdered gypsum is heated to approximately 170⁰C, it loses three-quarters of its water of crystallisation and becomes Hemi

Hydrate Gypsum Plaster (Class A Plaster), better known as *Plaster of Paris*. It has a very rapid setting time, which means it is difficult to trowel to a high finish, but it is suitable for repair work or for manufacturing plaster mouldings.

Self-assessment question 19.1

What would the chemical formula be for Hemi Hydrate Gypsum Plaster?

To prolong the setting time of the plaster and thus allow a longer period for trowelling smooth, an animal protein, such as *keratin,* may be added to Class A plasters to make them into Retarded Hemi Hydrate Gypsum Plaster (Class B Plaster), known as Thistle Plaster. These plasters normally come pre-mixed with the aggregate and are produced in two types:

- Type A – undercoat plasters, either *Browning* quality or metal lathing quality
- Type B – final coat plasters, either finish plaster, multi finish plaster, or board finish plaster.

However, when the set does take place with Class B plasters it does so very quickly and so a very high quality finish may not always be obtainable.

If the powdered gypsum is heated to 260°C it loses nearly all its combined water and becomes Anhydrous Gypsum Plaster (Class C Plaster), known as Sirapite Plaster. Approximately 1 per cent *alum* or zinc is added to accelerate the set. It benefits from a slower setting time than Class B plasters and is therefore useful as a finish coat, which is the only quality available.

Self-assessment question 19.2

What constituent in gypsum plasters appears to affect their setting time?

A harder plaster may be produced by heating the powdered gypsum to a temperature of 370°C. This produces a Superheated Anhydrous Gypsum Plaster (Class D Plaster), known as Keene's Cement. Again this is a finishing plaster only, but because of its extremely hard surface finish it can be specified for areas where there is a likelihood of the surface receiving knocks or abrasions.

Lightweight plasters are basically Class B plasters incorporating lightweight aggregates such as expanded perlite or exfoliated vermiculite. They have a density of approximately one-third that of plasters that are mixed with sand aggregates, and have better thermal insulation, sound insulation, fire resistance and condensation absorption characteristics than ordinary plasters. They also have superior adhesion properties to all backgrounds, but tend to be slower in drying than ordinary plasters with sand aggregates.

Self-assessment question 19.3

Why do lightweight plasters take longer to dry than ordinary plasters that are mixed with sand aggregates?

There are also a range of specialist plasters to suit most applications. These are:

- an undercoat plaster with good impact resistance
- an undercoat plaster for use on low suction backgrounds
- an undercoat plaster with superior impact resistance
- a cement-based plaster for use on walls following the installation of a new damp proof course
- an undercoat plaster for general re-plastering applications
- a finish plaster for use on plasterboard backgrounds
- a one coat plaster that can be applied up to 50mm in thickness
- a one coat plaster suitable for application by a plaster projection machine
- an undercoat plaster for use on surfaces where protection from X-rays is required.

Expanded metal lathing

In order to gain a good key for plastering on smooth surfaces, such as cast in situ concrete walls, it is sometimes necessary to fix a mechanical bonding aid. Expanded metal lathing (EML) consists of 24 gauge galvanised sheet steel, cut and expanded to form a diamond mesh and obtainable in sheets 450–600mm wide and 2.5–3.5m long. It is manufactured to the requirements laid down in *BS EN 13658-1:2005 Metal lath and beads. Definitions, requirements and test methods. Internal plastering*.

The sheets are fixed to the background by galvanised nails or by shot firing. EML is also useful for bridging across changes in background material, particularly where materials of differing thermal coefficients are used, to aid in the prevention of cracking to the plaster or render finish.

External angle beads are also manufactured from EML and are used at arrises to provide reinforcement at points liable to frequent impact damage.

Application of plasters

The number and thickness of coats depends on the type of background the plaster is being applied to and its level of 'trueness'. The application of plaster is covered by *BS EN 13914-2:2005 Design, preparation and application of external rendering and internal plastering. Design considerations and essential principles for internal plastering*. Most solid backgrounds can be covered with two coats with an overall thickness of 13mm, but where the surface of the background is irregular three coats may be required, having an overall thickness of 19mm.

For plasterboard and fibreboard a single coat of 2mm can be applied or, where a higher standard of finish is required, two coats having an overall thickness of 5mm may be specified. On metal lathing two coats are usually applied, with a thickness of 13mm.

The finished plaster surface should not show a deviation from true greater than 3mm on a 2m straight edge.

The quality of the plastering to a surface depends to a large extent on its application. The background must be examined to ensure that it has sufficient strength to support the plaster. The porosity and suction characteristics of the background will affect the adhesion of the plaster and its final strength. The absorption characteristics may need to be evened out by the use of artificial keys such as polyvinyl acetate emulsion (PVAC) to *BS 5270-1:1989 Bonding agents for use with gypsum plasters and cement. Specification for polyvinyl acetate (PVAC) emulsion bonding agents for internal use with gypsum building plasters*, which may develop better adhesion.

Self-assessment question 19.4

Give an example of a background having high porosity and suction characteristics.

In addition, the background should not contain high quantities of soluble salts that may cause efflorescence or even dampness on the finished plaster surface.

Self-assessment question 19.5

Give an example of a background that may contain high quantities of soluble salts.

A particularly important consideration is that of moisture and thermal movements occurring in the plaster. These may be caused by differential movement between the plaster and its background or relative movement at the junction of dissimilar materials. A common form of cracking is when a plaster that has high thermal expansion is applied to a background having low thermal expansion.

Self-assessment question 19.6

Give an example of a background that may have a low thermal expansion coefficient.

Where plastering is to be continued over mixed backgrounds a strip of EML fixed across the junction will minimise cracking. Alternatively, a straight cut through the plaster along the junction line may be adopted.

Another important consideration is to ensure that undercoats and final coats are compatible, since a common defect is often the bond failure between the two coats.

Plasterboard

Plasterboard comprises a solid gypsum plaster core encased in, and bonded to, specially prepared stout paper liners. Additives are used to help the plaster flow on to the paper liner during production and to improve the bond between the plaster and the paper liners. In addition further additives may be used to provide enhanced performance to specialist boards, such as thermal resistance, moisture resistance and fire resistance. *BS EN 520:2004 + Amendment 1:2009 Gypsum plasterboards. Definitions, requirements and test methods* classifies three main types of plasterboard:

- wallboard
- lath
- plank.

Wallboard is produced with one ivory and one brown coloured liner with square edges, suitable as a base for receiving a single coat of board finish plaster, or with tapered edges suitable for receiving taped and filled joints and one coat board finish plaster if desired. The plaster must be applied to the ivory coloured paper. On square edged boards, before the plaster coat is applied, the joints between the boards must be reinforced with a *jute scrim* strip 100mm wide, to prevent cracking occurring over the joints. Wallboard is available in a variety of sizes with thicknesses of 9.5mm, 12.5mm or 15mm. It has a thermal conductivity of 0.19W/mK. A special board is now produced that has a mass of $10kg/m^2$ to comply with the sound insulation requirements for internal and separating walls of Building Regulations Approved Document E, Sections 5.17 and 5.18.

Lath is similar to wallboard and is manufactured with one ivory and one brown coloured liner with square edges to receive a single coat of board finish plaster. This board is intended for ceiling finishes and is consequently narrower in width than wallboard, which makes for easier handling with thicknesses. It is supplied in 600mm or 900mm widths, which corresponds with normal joist spacing, 1220mm in length and thicknesses of 9.5 and 12.5mm.

Plank is similar to wallboard and is manufactured with one ivory and one brown face. It is supplied with tapered edges or square edges. It is the thickest of all the plasterboards, having a thickness of 19mm, and is useful in situations where additional strength, sound insulation or fire protection is required. It is 600mm wide and 2400mm in length.

In addition, wallboards are available with a vapour control membrane for use in applications where interstitial condensation needs to be prevented. Wallboards are also available with a backing of low density polystyrene, used in applications where the external wall construction needs an upgrade in thermal insulation. Other special boards are manufactured for specialist applications. These include:

- Fire resistant – plasterboard with glass fibre and other additives in the core to give increased fire resistance. This board is also available with a vapour control membrane.
- Acoustic – plasterboard with a higher density core for use in applications where greater levels of sound insulation are required.
- Impact resistant – plasterboard with heavy-duty paper facings and a higher density core to give greater impact resistance in heavy use areas.
- Moisture resistant – plasterboard with a silicone additive in the core and encased in water repellent liners. This is suitable as a base for ceramic tiling in wet use areas. This board is also available with enhanced fire resistance properties.
- Combined fire and impact resistant –a non-combustible glass reinforced plasterboard which combines fire protection properties with a high degree of impact resistance. This board is also available for providing fire protection to structural steel members.

Boards may also be manufactured faced with a white PVC film for ease of cleaning or with a perforated pattern to improve sound absorption characteristics.

Plasterboards are normally fixed to timber joists or studs by means of galvanised steel lath nails, the heads of which are filled after fixing or screwed into metal studs. Alternatively, the boards may be fixed direct to solid wall surfaces as dry lining by bonding to plaster dots or specially developed adhesive. The subject of dry lining will be considered in more detail in Chapter 21 (internal finishes). Plasterboards also form the main component in many lightweight partitioning systems. These will be considered in more detail in the next chapter (internal walls and partitions).

Sustainability

The gypsum for plaster and plasterboard has to be mined, which can cause damage to the environment and deface the landscape. In addition energy is required to crush the gypsum and to heat it to produce plaster. The manufacture of plasterboard also uses heat to dry the board once it has been produced. However, one of the greatest environmental impacts related to plasterboard is the amount of waste sent to landfill sites. This waste can create toxicity in both the land and associated water courses. Manufacturers of plasterboard are aware of the waste problem and are taking active steps to reduce it by 50 per cent.

Approximately 25 per cent of the content of plasterboard comes from recycled sources and gypsum is now being supplied from coal fired power stations through their flue gas desulphurization process. This adds additional benefits by reducing the amount of SO_2 being emitted into the air. Also, a large amount of the paper used for plasterboard lining comes from recycled sources. In addition, transportation energy is lower than would be the case if it was imported.

Plasterboard is seen as a very beneficial material within the construction industry. It is relatively inexpensive, provides high quality internal finishes to walls and ceilings and can enhance the thermal insulation, fire resistance and sound resistance of internal walls and ceilings.

Visit the companion website to test your understanding of Chapter 19 with a multiple choice questionnaire, and to see the following features illustrated with full-colour photos:

- **19.1** A sheet of wallboard
- **19.2** A skim coat of undercoat plaster to a wall prior to applying dry lining
- **19.3** An external angle bead to a plaster wall finish
- **19.4** Expanded metal lathing
- **19.5** Plaster finish to plasterboard dry lining

CHAPTER 20

Internal walls and partitions

Introduction

The primary function of an internal wall is to enclose or divide space. Generally such walls are used as partitions to form rooms in the house.

Internal walls may be constructed from a variety of materials and may be classified as either solid or lightweight. Solid partitions are normally constructed from bricks or blocks and may be loadbearing. Lightweight partitions are normally constructed from a framework of timber or light galvanised steel members with plasterboard facings, known as *stud partitions*. Generally lightweight partitions are non-loadbearing.

This chapter will consider the functional requirements of internal walls and partitions and the construction of solid and lightweight partitions for houses.

Functional requirements

Internal walls and partitions may be required to satisfy the following functional requirements:

- strength and stability
- fire resistance
- sound insulation.

The walls separating properties in semi-detached or terrace housing (often referred to as *party walls* or *separating walls*) require special attention in their design and construction, especially with regard to sound insulation and fire resistance.

Strength and stability

The internal wall may be required to carry loads and act as a support to the upper floor or roof construction in order to reduce the span of the joists or purlins. Where the internal wall is required to be loadbearing it should not be too thin to withstand the stresses set up by the applied load.

The thickness of loadbearing partitions will be dependent on their height and length. The Building Regulations Approved Document A, Section 2C10 stipulates that internal loadbearing walls of brickwork or blockwork should have a minimum thickness of 90mm where the height and length of the wall do not exceed 9m, and 140mm where the height and length of the wall exceed 9m but do not exceed 12m. However, similar to external loadbearing walls, the thickness of internal loadbearing walls should also be at least 1/16th of the storey height for adequate stability.

Self-assessment question 20.1

A non-loadbearing partition is 2.4m in height and 3.6m in length. What should be its minimum thickness?

The strength and stability of non-loadbearing partitions needs to be adequate enough to resist the accidental impact of persons or objects on their surfaces or the slamming of a door that might be included within their construction. These stresses are normally resisted within the framework construction of the partition.

Fire resistance

The internal wall or partition should possess adequate fire resistance to prevent fire from entering the room or to prevent the spread of fire between rooms. Loadbearing walls should have a minimum half-hour fire resistance. The materials used in the partition construction should therefore have low combustibility. This is not a problem with solid partitions built from bricks or blocks, but it could be a problem with lightweight partitions using timber or light steel framing.

Furthermore, the materials used for the finishes to the partitions will need to be carefully selected with regard to their surface spread of flame performance. Generally plasterboard finishes are used for lightweight partition systems and this material has satisfactory inherent fire resistance and good resistance to the surface spread of flame.

Where high fire resistance is required, specialised fire resisting plasterboard can be used in conjunction with the installation of incombustible quilt within the cavity of stud partitions.

The Building Regulations Approved Document B, Section 5.3 requires that separating walls be designed and constructed as compartment walls for fire resistance purposes. They need to run the full height of the building (including inside the roof space) and should not be greater than 20m in height. The separating wall needs to be constructed to within 25mm of the top of the roof trusses, and soft packing, such as mineral wool as fire stopping, should be used between the top of the wall and the roof underlay.

Furthermore, the separating wall should be extended through the roof for a height of at least 200mm above a roof covering of low combustibility or a zone of the roof at least 1500mm wide on either side of the separating wall should have a covering of low combustibility.

Sound insulation

The internal wall or partition should possess adequate sound insulation to prevent the transmission of sound between rooms. This is particularly important in walls separating bathrooms and living room, dining room, study or bedroom where the partition should have a sound reduction of 45dB over the frequency range of 100–3150Hz. The materials used in the wall construction should either be of adequate thickness and density, which is achievable with solid partitions, constructed of materials having a minimum density of 600kg/m^3, or sound absorbent quilt may need to be installed within the cavity of stud partitions, with high density plasterboard facings.

The Building Regulations require sound insulation testing to demonstrate compliance and this should comprise two individual airborne tests: one to test the airborne sound transmission between one pair of reception rooms situated on opposite sides of the partition and another test measuring the airborne sound transmission between two bedrooms situated on opposite sides of the partition.

Care also needs to be taken where the partition joins with other elements within the building, such as floors, roofs, external walls and other internal walls, in order to control flanking transmission, that is structure borne sound transmission between the connecting elements of the structure.

In order to provide effective sound insulation between adjoining properties, the *Building Regulations Approved Document E, Section 2: Separating walls and associated flanking constructions for new buildings* gives recommendations for the design of four types of wall construction:

- solid masonry wall
- cavity masonry wall

- masonry wall between independent panels
- framed walls with absorbent material.

Solid masonry wall

In this type of separating wall the resistance to airborne sound transmission is mainly influenced by the mass per unit area of the wall.

This could be constructed from brick with plaster or plasterboard to both sides, providing an overall minimum mass to the wall of 375kg/m^2. Alternatively the wall could also be constructed from concrete block with plaster or plasterboard both sides, providing an overall minimum mass to the wall of 415kg/m^2.

Self-assessment question 20.4

Why should the Building Regulations requirements for solid masonry separating walls stipulate an overall minimum mass to the wall?

Cavity masonry wall

The resistance to airborne sound transmission of this type of separating wall depends on the mass per unit area of the two leaves and the degree of isolation achieved between them. This isolation is greatly affected by the type of wall ties used. As discussed previously in Chapter 8 (external walls), the butterfly type of wall tie is favoured over the double triangle or vertical twist type of ties for sound insulation purposes.

This type of separating wall could be constructed from leaves of dense concrete block with plaster or plasterboard to both room faces and a minimum 50mm cavity providing an overall minimum mass of 415kg/m^2. Alternatively the wall could be constructed from two leaves of lightweight aggregate concrete block with plaster to both room faces and a minimum cavity width of 75mm, providing an overall minimum mass of 300kg/m^2 or with plasterboard to both room faces, a 75mm cavity and a minimum mass of 275kg/m^2. Alternatively both leaves of the wall could be built with aircrete blocks with two sheets of plasterboard to both room faces with joints staggered, a minimum cavity width of 75mm, providing an overall minimum mass of 150kg/m^2.

Self-assessment question 20.5

Why should the cavity for a separating wall built with two leaves of lightweight aggregate blockwork be wider than the cavity for a separating wall built with two leaves of concrete blockwork with dense aggregate?

Masonry wall between independent panels

The resistance to airborne sound relies substantially on the masonry material used in the core, in particular its mass per unit area, but also the isolation and mass per unit area of the material used for the independent panels also has some influence.

The masonry walls to this construction could be a solid or cavity wall with independent panels of two sheets of plasterboard separated by a cellular core situated on both sides but not in contact with the masonry core. Suitable specifications would be a core of dense aggregate concrete block with independent panels either side providing an overall minimum mass of 300kg/m^2 or a lightweight concrete block core with independent panels of composite plasterboard sheets with a cellular cardboard central core either side.

There should be a minimum gap of 35mm between the independent panels and the masonry core if the panels are not supported on a frame. If they are supported on a frame then the minimum gap between the independent panels and the masonry core can be reduced to 10mm.

Self-assessment question 20.6

Why should a minimum 35mm cavity be provided between each independent panel and the masonry wall?

Framed walls with absorbent materials

In this type of separating wall the resistance to airborne sound transmission is dependent on the mass per unit area of the leaves, the isolation of the frames and the amount of sound absorbent material in the cavity between the frames.

This construction may consist of timber or steel frames of 200mm overall thickness with sound absorbent material such as unfaced mineral fibre batts or quilt 25mm thick if suspended in the cavity between the frames or 50mm thick if fixed to one frame and having a minimum density of 10kg/m^3, lined by two or more layers of plasterboard having a combined thickness of 30mm to each side. Alternatively the timber frames may be attached to a masonry core.

There are also requirements within the Building Regulations Approved Document E, Section 2 for the construction of the junctions between separating walls and the external wall, the roof and the floors of these properties.

Solid partitions

These are normally loadbearing and possess good fire resistance and sound insulation performance. They are usually constructed from bricks or blocks.

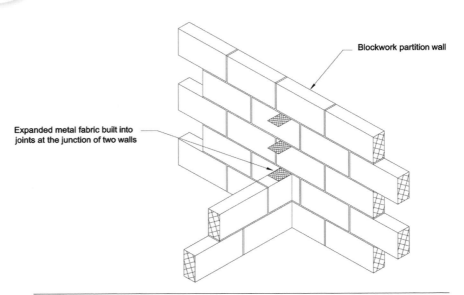

Blockwork partition wall

Expanded metal fabric built into joints at the junction of two walls

Figure 20.1 Bonding in partition walls using a strip of expanded metal fabric

Where bricks are used the wall will normally be half brick in thickness and constructed in stretcher bond. The bricks should have a minimum compressive strength of $5N/mm^2$. Because finishes are likely to be applied to these walls common bricks may be used. Internal quality bricks are available that have grooved surfaces to provide a key for plastering.

Where blocks are used they may be manufactured from either clay or concrete. The latter may be solid, cellular or hollow in construction. The blocks should have a minimum compressive strength of $2.8N/mm^2$.

Where walls are bonded at their intersection, blocks may be built in or a strip of expanded metal fabric may be incorporated at the junction (see Figure 20.1).

Loadbearing partitions require adequate foundations to transfer their loads to the ground beneath. These may be independent strip foundations or, alternatively where the loadbearing capacity of the subsoil near to the surface of the ground is adequate, the ground floor slab may be thickened beneath the partition (see Figure 20.2). Blocks used below DPC level should be suitable for that situation. The DPC should be linked with the DPM in the floor construction whether the solid partition is loadbearing or not (see Figure 20.2).

Where solid partitions are built to upper storeys of the house it is useful to ensure that their location coincides with those on the lower storey in order to transfer their loads directly to the foundations. Where it is not possible to locate the partitions to coincide with those in the lower storey, a suitable support for the upper partition must be provided. This should ideally be a RSJ rather than a timber bearer or joist (see Figure 20.3).

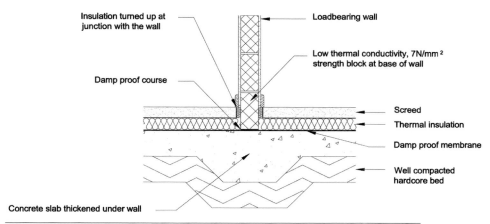

Figure 20.2 Thickening of a ground floor slab beneath a loadbearing partition

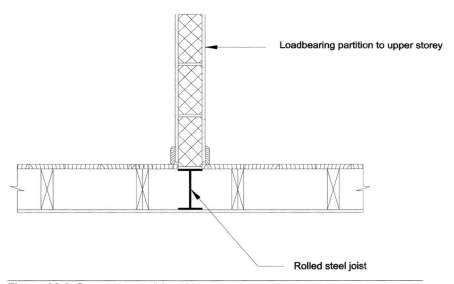

Figure 20.3 Support to a solid partition to an upper storey

Self-assessment question 20.7

Why is it not ideal to support upper floor solid partitions on a timber joist or bearer?

Lightweight partitions

These are generally non-loadbearing. Consequently they may be constructed of lightweight materials to reduce the dead load on the structure and foundations. Thus they may be constructed anywhere in the house without requiring special foundations or bearings for support. However it is beneficial if building the partition off a concrete ground floor slab to provide a DPC immediately beneath the partition to prevent any residual moisture from the concrete slab affecting the timber or steel within the partition.

Lightweight partitions are generally stud partitions with a lightweight steel or timber framework and plasterboard facings to both sides.

Timber stud partitions

These consist of a 63 x 38mm timber framing comprising a sill or bottom member, fixed to the floor, a head member fixed to the ceiling and vertical *studs* at 450–600mm centres fixed between the head and sill members and strengthened by horizontal *noggings* (see Figure 20.4).

Door openings may be accommodated within the partition by framing with stud, head and possibly sill members. Coverings may be of any suitable sheet material such as plywood, particleboard, oriented strand board, flaxboard or fibreboard, but the most popular material is plasterboard, generally wallboard with either a skim coat plaster finish or decorated directly.

Metal stud partitions

A proprietary metal stud partition is also available. This utilises 70 × 34mm deep, light galvanised steel channel sections similar to those used in steel frame house construction (see Figure 20.5). This is lighter than the timber partition and installation costs are claimed to be lower than those for the timber partition and it does not suffer from moisture movement.

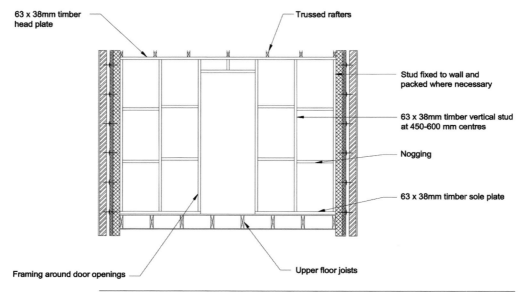

63 x 38mm timber head plate

Trussed rafters

Stud fixed to wall and packed where necessary

63 x 38mm timber vertical stud at 450-600 mm centres

Nogging

63 x 38mm timber sole plate

Framing around door openings

Upper floor joists

Figure 20.4 A timber stud partition

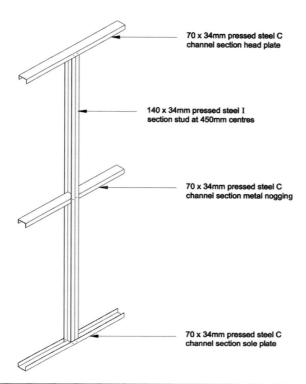

70 x 34mm pressed steel C channel section head plate

140 x 34mm pressed steel I section stud at 450mm centres

70 x 34mm pressed steel C channel section metal nogging

70 x 34mm pressed steel C channel section sole plate

Figure 20.5 A steel stud partition

These partitions have therefore become more popular than their timber framed counterparts in recent years.

The hollow construction of the stud partition makes it easier to accommodate thermal or sound insulation materials as well as service conduits, pipes and cables.

Visit the companion website to test your understanding of Chapter 20 with a multiple choice questionnaire, and to see 17 full-colour photos illustrating this chapter.

CHAPTER 21

Internal finishes

Introduction

Upon completion of the structure, suitable finishes are required to the internal surfaces of the house. Surfaces requiring special finishes are:

- floors
- walls
- ceilings.

Suitable finishes for these surfaces are dependent on specific functional requirements, which are determined by the surface being considered and its location.

This chapter will consider the functional requirements of internal finishes and then consider suitable floor, wall and ceiling finishes for houses. Particular attention will be paid to floor screeds, their types and the method of laying them, as well as the technique of applying dry lining as a wall finish.

Functional requirements

These may be considered under the following headings:

- durability
- comfort and safety
- physical attributes
- maintenance.

It is important to bear in mind that not all the factors listed under these headings will apply in every case. For instance the floor finish to a kitchen may well require different functional requirements than the wall finish to a bathroom or the ceiling finish to a living room. However, those functional requirements that are appropriate to a particular situation will need to be considered when selecting the most appropriate finish for the situation.

As this book is only concentrating on domestic construction, some functional requirements that may be appropriate for consideration when selecting finishes to other types of buildings will not be considered here.

Durability

The following factors need to be taken into account:

- Resistance to abrasion – damage may be caused by objects rubbing against the surface.
- Resistance to water or other liquids – damage may be caused if the surface finish is not waterproof in areas where water or other liquids may be frequently used or spilled.
- Resistance to indentation – damage caused by objects banging into the surface finish or the damage caused by high heels on soft floor finishes.
- Resistance to impact – damage caused by objects banging into the surface finish. This may cause cracking to fragile or brittle finishes.
- Resistance to sunlight – ultra-violet degradation causing embrittlement of some materials or fading of colours in other materials.
- Resistance to moulds and fungi – mould growth may be a problem on finishes where humidity is high and ventilation is poor allowing surface condensation to develop.
- Resistance to high temperatures or fire – some finishes may need to have a low rating for the spread of flame across their surface in order to ensure that a fire is not able to spread easily from one room to another in a house.
- Resistance to splitting – caused by wear and tear of the finish.
- Resistance to tearing – caused by wear and tear of the finish.
- Resistance to cracking – caused by wear and tear of the finish or created by stresses and strains on the finish by the reaction between the finish and the background or items attached to the surface.
- Amount of moisture movement – compatibility with the background and surrounding materials.
- Amount of thermal movement – compatibility with the background and surrounding materials.
- Strength in bending – particularly important where the finish is likely to be attached to a curved surface or one in which bending due to deflection under load may occur.
- Strength in compression – particularly important where the finish is attached to a surface which is likely to be subjected to compression forces.
- Strength in tension – particularly important where the finish is attached to a surface which is likely to be subjected to tension forces.
- Strength in shear – particularly important where the finish is attached to a surface which is likely to be subjected to shear forces.

Comfort and safety

The following factors need to be taken into account:

- Freedom from slipperiness – particularly important for floor finishes where liquid spillages might occur.

- Warmth – contribution to the thermal performance of the element to which the finish is attached. Timber finishes are warmer than stone, concrete or clay tile finishes.
- Quietness – contribution to the acoustic performance of the element to which the finish is attached. Carpets are soft and therefore quiet; tiles are hard and can be noisy when walked upon.
- Resilience – ability of the finish to spring back into shape after being walked upon.
- Sound absorption and transmission – soft finishes can contribute to a reduction in the transmission of sound energy through an element if they are applied to a surface. Echoes are reduced in a room when soft finishes are used. Hard floor finishes can contribute to an increase in impact sound.
- Light reflection and transmission – finishes that are light in colour or shiny in texture are likely to provide greater light reflection than dark or matt surfaces. Similarly transparent surfaces will provide greater light transmission than will opaque surfaces.
- Thermal conductance – some materials used for surface finishes, such as metals, may conduct heat more readily than others.
- Toxicity – surface finishes used in areas where food is prepared should not be toxic or become toxic if moistened.
- Moisture absorption – surface finishes that can absorb small amounts of moisture, such as plasters, can be helpful in reducing the incidence of surface condensation in areas of high humidity such as bathrooms and kitchens. However, timber products may swell excessively if they are used in areas where moisture absorption is likely to be high.
- Fire spread – the choice of materials for internal finishes can affect the spread of a fire within a building, even though they may not be the source of the fire itself. The materials for finishes need to be assessed for their ease of ignition, the rate at which they will emit heat when burning and the degree to which they may emit smoke and fumes when burning. These factors are particularly important when selecting finishes in areas that are likely to be the main escape routes from the building.

Physical attributes

The following factors need to be taken into account:

- Appearance – colour, design and texture of the finish material.
- Shape – this has an effect on the appearance of the finish, but can also affect how easily the finish can be attached to the background.
- Surface texture – a smooth surface is easily cleaned whereas a textured surface can trap dirt. However any unevenness in a smooth surface may be highlighted when it is illuminated.
- Dimensions – large units are quicker to fix than smaller units but may require more cutting if fixed in small areas or installed around fixed objects.

- Weight – this has an effect on the dead load of the building but may also have a significant effect on the attachment to the background. Weak backgrounds will not be able to support heavy finishes.
- Method of fixing – this could be by mortars, adhesives or mechanical fixings dependent on the type of finish.
- Cost – expensive finishes may convey a sense of opulence to the observer. Finishes having a low initial cost may need to be replaced more frequently than more expensive finishes.
- Ease of cutting – particularly important where finishes need to be installed around fixtures and fittings.
- Ease of handling – large or bulky materials may be more difficult to handle than smaller and lighter materials.
- Ease of jointing – this includes the method of jointing finishes to each other and the method of jointing finishes to other materials in the building. Joints must be able to satisfy the functional requirements of the finishes they join and may also need to accommodate tolerances for moisture and thermal movements.

Maintenance

The following factors need to be taken into account:

- Ease of cleaning – some finishes may be easily cleaned by wiping with a damp cloth or sponge. Others may require special chemicals to clean them.
- Frequency of cleaning – some finishes may need more frequent cleaning than others due to their location or frequency of use. Frequent cleaning may be required to maintain standards of hygiene in food preparation areas.
- Ease of repair – some finishes may be easier to repair than others. Some finishes may require complete replacement rather than repair.
- Life expectancy – linked to life cycle costing. Some finishes may need to be replaced more frequently than others.

Self-assessment question 21.1

List the functional requirements that would need to be considered in the selection of finishes for the following areas:

A. Kitchen floor
B. Bathroom wall
C. Living room ceiling

Floor finishes

It is possible to provide a trowelled finish to solid concrete floors that is smooth and level enough for the application of most floor finishes.

However, it is seldom practised because following trades could cause damage to the surface of the floor before the finish is applied, meaning that the surface of the floor will need to be re-smoothed and re-levelled prior to the application of the finish. In addition, by providing a screed to the surface of the floor, conduits, pipes and cables for services may be accommodated within the thickness of the screed.

Floor screeds

The screed must be laid correctly to provide a sound and level surface for the final floor finish and to prevent the subsequent cracking and lifting of the screed. Screeds are generally produced from dense sand:cement mixes (proportioned 1:4 by weight). Alternatively lightweight aggregates may be substituted for the sand. This is particularly useful where the screeds are likely to be thick and the dead loads imposed on the floor are an important consideration.

The amount of water included in the screed mix is also important. The mix should contain just enough water to allow the screed mixture to be moulded in the hand without squeezing out any excess water. Lightweight aggregates absorb more water than dense aggregates; therefore they take longer to dry out. The drying time of screeds can be critical, particularly where moisture sensitive floor finishes are to be laid over them. The *Building Research Establishment Good Building Guide 28 Part 2: Concrete floors, screeds and finishes* recommends, in general, a drying time of one month per 25mm of screed thickness should be allowed for natural drying. This drying time can be accelerated with the use of dehumidifiers.

Screeds need to be compacted well when they are laid, otherwise hollow spots may occur in the finished screed and its durability and performance will be affected.

There are two main types of floor screed:

* bonded – where the screed is bonded to the floor to ensure adequate adhesion is achieved
* unbonded – where the screed is held down by its own mass.

Bonded screeds

These may be of two types:

* monolithic screed
* separate construction.

Monolithic screed

This is laid on an in situ concrete base before it has set, thus minimising shrinkage. The screed can be laid to a minimum thickness of 12mm (see Figure 21.1) and is generally used where a special concrete mix for the finish to the concrete is required, such as *granolithic* or *terrazzo*, and the cost of providing this specification to the entire concrete slab would be exorbitant and unnecessary.

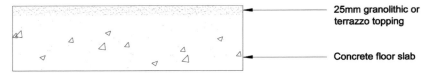

Figure 21.1 A monolithic screed

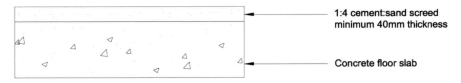

Figure 21.2 A separate screed

Separate construction

This is laid on a hardened concrete base to a minimum thickness of 40mm (see Figure 21.2). The strength of the bond between the screed and the base is dependent on the roughness of the surface finish to the concrete slab. Where the slab has received a tamped surface finish this is normally sufficient to provide a good key for the screed. Alternatively a key can be produced mechanically by means of hacking the concrete surface or applying a bonding agent such as PVAC adhesive. The concrete surface must be dampened before the application of the screed to reduce suction.

Self-assessment question 21.2

Why does the surface of the concrete slab need to be dampened to reduce suction?

Unbonded screeds

These may be of two types:

- unbonded construction
- floating screed.

Unbonded construction

This screed is used where the bond between the concrete base and the screed has been broken.

Self-assessment question 21.3

Give an example of a situation in which the bond between the concrete base and the screed is broken.

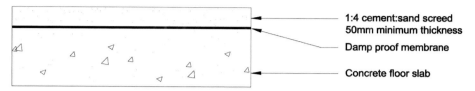

Figure 21.3 An unbonded screed

Because there is no bond between the screed and the slab, the screed must rely on its own mass to prevent it from lifting. The minimum thickness of this screed must therefore be 50mm (see Figure 21.3).

Floating screed

In order to achieve satisfactory thermal or acoustic insulation values, the floor finishes are occasionally separated from the concrete base by a layer of insulating material. To prevent the insulation from becoming wet, it is normally covered with a damp proof membrane and a layer of chicken wire or reinforcing fibres is embedded in the screed to strengthen it. This screed has a minimum thickness of 65mm (see Figure 21.4).

Self-assessment question 21.4

Why does the floating screed require a layer of chicken wire or reinforcing fibres embedded in the screed to strengthen it?

Where improved sound insulation to upper floors is required (generally in blocks of flats) a floating screed on a concrete floor is often an ideal solution. Where floating screeds are used internal walls and partitions must be constructed off the structural floor.

Self-assessment question 21.5

Why do internal walls have to be built off the structural floor where floating screeds are used?

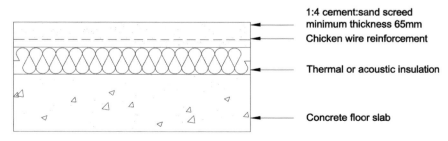

Figure 21.4 A floating screed

Floor coverings

Floor coverings may be divided into four distinct groups:

- in situ coverings
- sheet coverings
- tile coverings
- timber and timber product coverings.

In situ coverings

These are generally monolithic screed finishes, such as granolithic flooring, in which the screed contains a crushed granite aggregate or terrazzo, in which the screed contains decorative marble aggregates. The surface of this can be polished to a fine finish where the marble aggregates enhance the appearance of the floor finish. Both finishes provide a hardwearing surface. The granolithic floor can be used for a garage.

Mastic asphalt can be used as a floor covering. It is hard wearing but can suffer from damage due to indentation from heavy furniture. It can become slippery when wet, but has very good damp resistance and can be used as a damp proof membrane in older properties where the ground floor slab does not contain a DPM. It has a limited colour range and is therefore rarely used on new house construction.

Sheet coverings

A variety of sheet materials are available, including:

- Linoleum – made by pressing a plastic material comprising oils, cork and other fillers and pigments onto a *jute* canvas or glass fibre and polyester backing to *BS 6826:1987 Specification for linoleum cork carpet sheet and tiles*. The sheet can be supplied in plain, printed or inlaid finishes. Its sustainability credentials are good. It is durable, non-toxic, recyclable and biodegradable.
- Flexible PVC – polyvinylchloride is mixed with fillers, extenders, pigments and stabilisers. The mixture is *calendered* between heated rollers to produce a sheet of the correct thickness to *BS EN 649:2011 Resilient floor coverings. Homogenous and hetrogenous polyvinylchloride floor coverings. Specification*. Designs can be printed onto the sheet and then a surface of clear plastic can be laminated on top to seal and protect the design. The sheet can be 'welded' together at joints if required. Where this type of floor covering is used on concrete ground floor slabs, the floor should contain a damp proof membrane. Its sustainability credentials are quite good. Although there is a petrochemical content and there are concerns over the toxins released in the manufacture of PVC, it is recyclable and durable.
- Rubber – natural or synthetic rubber is mixed with varying filling compounds and pigments to produce different textures and colours to *BS 1711:1975 Specification for solid rubber flooring*. Sulphur is added

to *vulcanize* the rubber. This floor covering has good resistance to oils, fats and greases and is antistatic. However it is rarely specified for houses and tends to be more suitable for buildings where the performance characteristics of the covering far outweigh its appearance, which tends to be rather bland in comparison with other sheet coverings. The sustainability credentials of natural rubber are good. It is recyclable, non-toxic, durable and is derived from a natural resource which is renewable. However if it is destroyed by incineration, toxic fumes are emitted. The sustainability credentials of synthetic rubber are similar to those of natural rubber. However it is non-biodegradable and may contain PVC or plasticisers.

- Carpet – a variety of fibres and different methods of manufacture provide a diversity of floor coverings of varying quality and performance characteristics. The fibres used in carpet manufacture may be wool, synthetic materials or a mixture of both. The *yarn* produced can be heated to produce a hardwearing twist *pile* or woven into a soft velvet *pile*. Needle pile floor coverings are covered by *BS EN 13297:2000 Textile floor coverings. Classification of needle pile floor coverings*. Carpets are manufactured by weaving or stitching the yarn into the backing material. Alternatively, with synthetic fibres, a bunch of fibres can be needle punched into the backing or glued to the backing. The backing may be jute or, more commonly, *latex*. The number of tufts per unit area determines the density of the carpet. Dense carpets with a short pile tend to be harder wearing than less dense carpets with a longer pile. Carpets are available in a wide range of colours, patterns and textures. Where the carpet is not supplied with a latex backing it is necessary to provide an *underlay* of sheet rubber or polyurethane foam. Underlays are covered by *BS EN 14499:2004 Textile floor coverings. Minimum requirements for carpet underlays*. The sustainability credentials of carpets depend to a great extent on the materials of their manufacture. Nylon carpets have good wear and resistance to abrasion performance and are recyclable. However they are manufactured from petrochemical polymers and are non-biodegradable. Wool carpets, on the other hand, are natural products from renewable resources. They have good fire resistance and are biodegradable. However they are not suitable in areas of high usage. Polyester and polypropylene have similar sustainability credentials to those of nylon, although recycled polyester is now becoming widely available. The fibres are produced from recycled soft drinks bottles and thus have good sustainability credentials.
- Laminate – a multi-layer synthetic material comprising an inner core of high density fibreboard with an impregnated paper backing and a surface layer of laminate material with an applique design which simulates wood, ceramics or stone with a clear protective covering. The whole combination of materials is then fused together in a high pressure lamination process to form panels of 19–25mm in thickness. The panels are tongued and grooved on all edges.

Tile coverings

A variety of tiles are available, including:

- Clay – these are available in three types:
  - Quarry tiles – manufactured from unrefined clays to *BS EN 14411:2006 Ceramic tiles: Definitions, classification, characteristics and marking*. They are of less uniform composition than vitrified ceramic tiles. They are classified according to their water absorption: Class 1 tiles have a maximum water absorption of 6 per cent, whilst Class 2 tiles have a maximum water absorption of 10 per cent. They are laid and jointed in cement:sand mortar and expansion joints of compressible material need to be laid around the perimeter of the tile covering. They have a limited range of colours, mainly reds, browns, and buffs, but they have very good wearing qualities.
  - Ceramic – manufactured from refined clays with *fluxes* to increase *vitrification* to *BS EN 14411*. They are available as either vitrified, having a maximum water absorption of 4 per cent or fully vitrified, having a maximum water absorption of 0.3 per cent. They are available in a much wider range of colours, textures and designs than quarry tiles and are also thinner and have a fine, smooth surface finish.
  - Ceramic mosaics – supplied as 38mm square pieces mounted on a paper backing in a pattern ready for laying.

    The sustainability credentials of clay tiles are mixed. They are recyclable and recycled materials can be used in their manufacture. They are made from an abundant resource. However the extraction of the clay can destroy natural habitats and have a detrimental effect on the landscape. In addition the energy used to manufacture the tiles is quite high.

- Cork – formed of cork granules compressed and baked at high temperature so that the natural resins present in the grain bond the particles together to *BS EN 12104:2000 Resilient floor coverings. Cork floor tiles. Specification*. The material thus formed is then cut into tiles and these are fixed to the surface of the floor with adhesive. The tiles need to be sealed to provide good wearing characteristics and resistance to water. They have extremely good thermal resistance and therefore provide a floor covering that is warm to the touch.

  The sustainability credentials of cork are extremely high. It is a natural and renewable resource, which is durable, recyclable and biodegradable. It uses very little energy in production but is imported, so transportation energy is fairly high.

- Flexible PVC – as the sheet material previously considered.
- Carpet – as the sheet material previously considered.

Timber and timber product coverings

A variety of materials are available, including:

- Board and strip – wood strip flooring is available in widths up to 100mm. Wood board flooring is available in widths greater than

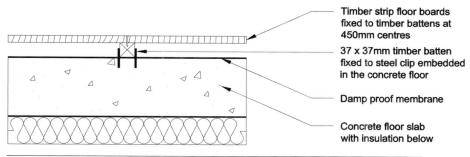

Timber strip floor boards fixed to timber battens at 450mm centres

37 x 37mm timber batten fixed to steel clip embedded in the concrete floor

Damp proof membrane

Concrete floor slab with insulation below

Figure 21.5 Timber strip flooring

100mm. Thicknesses are generally 20, 32 or 38mm. Although not true in every case, most strip flooring is made from hardwoods and board flooring is generally made from softwoods. The strips or boards are normally nailed to timber joists and fixed together by tongued and grooved joints. On concrete sub-floors it is possible to fix timber strip flooring to timber battens fixed to galvanised steel clips embedded in the floor (see Figure 21.5).

- Wood block – generally produced from hardwoods, flat or quarter sawn into blocks 90mm in width, 230 or 300mm in length and 19–38mm in thickness to *BS 1187:1959 Specification for wood blocks for floors*. They are fixed to the floor screed using bitumen emulsion. The blocks are normally laid in a herringbone or basket weave pattern.
- Parquet – manufactured from specially selected hardwoods and cut to form patterns either by laying the cut sections directly onto the sub-floor or pressed onto a backing and laid as panels. If the sections or panels are laid onto timber boarding then the boards should be covered with hardboard or plywood sheeting prior to laying the parquet covering to mask any surface irregularities in the floorboards.

The joint between the floor and wall finishes is normally covered by a skirting board. These are available in a variety of sizes and designs and are normally timber.

Timber has very high sustainability credentials. It is produced from a natural and renewable resource. It is durable, recyclable, biodegradable, non-toxic and uses very low amounts of energy in its conversion. However, tropical hardwoods are less sustainable than softwoods from temperate forests and incur high amounts of energy in their transportation.

Wall finishes

Although it is possible to provide an internal wall finish using facing bricks or building stone, it is not usual. Most internal finishes to walls comprise an application of plaster or plasterboard with a paint or wallpaper covering.

Where walls have been newly plastered it is desirable to finish them with a coat of emulsion paint for the first six months rather than cover them with an oil-based paint or wallpaper.

Self-assessment question 21.6

Why should newly plastered walls be decorated with emulsion paint, rather than oil-based paint or wallpaper, for the first six months?

Other wall finishes are tiling and panelling.

Wall tiling

Tiles are manufactured from special clays containing silica and alumina with sodium or potassium alkalis, mixed with a quantity of flint to resist *crazing* in the tile and limestone to reduce shrinkage of the tile during firing. The materials are then mixed with water to form a slurry and run to a storage tank. When required the slurry is pumped to presses where the water is extracted by pressure and the plastic clay left is then dried in stoves to reduce the moisture content to approximately 10 per cent. The dried clay is then crushed to dust of controlled particle size in grinding pans and the particles are formed into tiles in presses.

The formed tiles are then loaded onto trucks, which pass through a drier prior to entering a tunnel oven for firing. Once the 'biscuit' tile has cooled it can be glazed. The glaze is applied as a liquid and the biscuit tile, being very absorbent, soaks up the water leaving the glaze as a powdery crust on the surface of the tile. The tiles are then laid on the shelves of a fireclay truck and passed through another tunnel oven to fix the glaze to the surface of the tile.

There are two types of glaze, earthenware glazes and coloured enamels. Coloured enamels may have a glossy surface or an eggshell or matt surface.

Wall panelling

Traditionally wall panelling would be made from hardwood and formed into panels with stiles and rails similar to the technology used for the construction of solid timber panelled doors. Modern wall panelling generally consists of sheets of plywood or medium density fibreboard finished with a veneer and fixed to 100 x 19mm softwood framing, screwed to 50 x 25mm softwood *grounds* which in turn have been fixed to the backing wall. The panels may be fixed together by tongued and grooved joints on their edges (see Figure 21.6).

The panels are normally installed as *wainscoting* meaning that they extend from the floor to a height of approximately one metre above the floor level and are topped with a timber *dado rail*.

Dry lining

The use of plasterboard as a *dry lining* has become very popular in recent years for the following reasons:

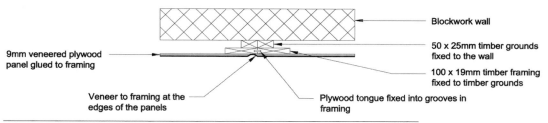

Figure 21.6 Timber wall panelling

- It provides a smooth finish to the wall that can receive direct decoration.
- It is a dry finish and therefore does not suffer from the problems of a prolonged drying period that besets wet plaster finishes.
- It can be applied by semi-skilled operatives, for less cost than conventional plaster finishes.
- It can incorporate thermal insulation materials and a vapour control layer to reduce the incidence of interstitial condensation in the wall.
- It is quick to apply.

However there are also some disadvantages to the use of dry lining, such as:

- It has limited strength for supporting heavy fixings, such as cupboards.
- It is susceptible to impact damage and therefore may not be suitable in areas where such damage may occur unless impact resistant plasterboard is used.

Traditionally the dry lining boards were fixed to preservative treated timber battens, fixed vertically to the backing wall and horizontally at ceiling and skirting levels, in a similar manner to that used in wall panelling. The spacing of the vertical battens is determined by the thickness of the plasterboard used, generally 450mm centres for 9.5mm thick boards and 600mm centres for 12.7mm boards (see Figure 21.7).

Thin galvanised metal channel sections, known as *furrings*, may be used instead of timber battens. These are bonded to the backing wall by adhesive. The metal furrings are positioned at similar intervals to the timber battens, but provide some extra latitude for unevenness in the background. The plasterboard is then screwed to the metal furrings (see Figure 21.8).

Self-assessment question 21.7

Why can the metal furrings provide a greater degree of latitude for unevenness in the background than can timber battens?

An alternative method for fixing plasterboard dry lining to the backing wall has been the 'dot and dab' method, where 75 x 50mm 'dots' of bitumen impregnated fibreboard were fixed with plaster to the backing wall at 1m centres vertically and 1.8m centres horizontally, to provide a true, flat ground. Intermediate dots were then placed to coincide with

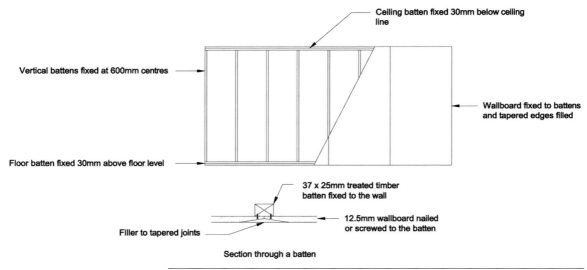

Ceiling batten fixed 30mm below ceiling line

Vertical battens fixed at 600mm centres

Wallboard fixed to battens and tapered edges filled

Floor batten fixed 30mm above floor level

37 x 25mm treated timber batten fixed to the wall

12.5mm wallboard nailed or screwed to the batten

Filler to tapered joints

Section through a batten

Figure 21.7 Dry lining using timber battens

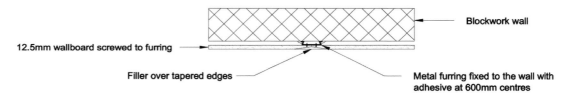

Blockwork wall

12.5mm wallboard screwed to furring

Filler over tapered edges

Metal furring fixed to the wall with adhesive at 600mm centres

Figure 21.8 Fixing of plasterboard dry lining to metal furrings

the board joints. Once these dots had set, thick 'dabs' of plaster were applied vertically between the dots, so that they stood proud of them and the plasterboard was then pressed into position on these dabs. A temporary nail fixing to the dots was achieved whilst the dabs were setting (see Figure 21.9).

The dot and dab method has now been replaced by the use of dabs of adhesive applied to the backing wall in vertical rows, with intermediate horizontal dabs applied at ceiling level. A continuous bead of adhesive is then applied at skirting level and the plasterboard is pressed in to the dabs to provide a level finish (see Figure 21.10).

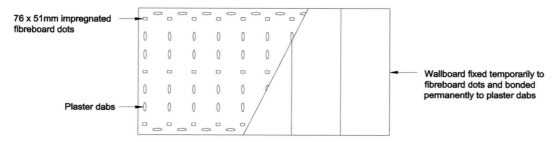

76 x 51mm impregnated fibreboard dots

Wallboard fixed temporarily to fibreboard dots and bonded permanently to plaster dabs

Plaster dabs

Figure 21.9 The dot and dab method of fixing dry lining

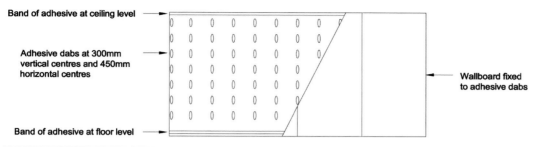

Band of adhesive at ceiling level

Adhesive dabs at 300mm vertical centres and 450mm horizontal centres

Band of adhesive at floor level

Wallboard fixed to adhesive dabs

Figure 21.10 Fixing of plasterboard dry lining with adhesive dabs

Ceiling finishes

Most modern ceilings are constructed using wallboard plasterboard fixed with lath nails to the underside of the floor or roof joists. The joints between the boards are covered with a perforated tape or *jute scrim* and then covered with filler to prevent cracking. The boards can be finished with a single coat of board finish plaster or decorated directly. The board or plaster finish may be painted or covered with wallpaper. Another popular finish to ceilings has been the application of Artex, a proprietary coating that can be applied directly to plasterboard without the need for a plaster finish, and can then be textured whilst wet with either a combed or stippled finish.

The joint between the wall and ceiling finish may be covered by coving. Traditionally this was produced from Class A plasters, reinforced with horsehair and cast in special moulds to a variety of designs. Modern coving tends to be manufactured from plasterboard and is fixed to the wall and ceiling surfaces with a special adhesive.

Visit the companion website to test your understanding of Chapter 21 with a multiple choice questionnaire, and to see the following features illustrated with full-colour photos:

- **21.1** Adhesive dabs applied to the inside of a pressed steel lintel to receive a plasterboard finish
- **21.2** Adhesive dabs for fixing dry lining to a solid background wall
- **21.3** Ceramic floor tiles
- **21.4** Fixing plasterboard to the underside of the ceiling ties of trussed rafters
- **21.5** Glazed ceramic wall tiles
- **21.5** Metal furrings awaiting fixing of plasterboards
- **21.6** Plasterboard ceiling awaiting taped finish to the joints
- **21.7** Skim plaster finish to plasterboard dry lining
- **21.8** Timber skirting to the junction between the floor and wall finish
- **21.9** Unglazed ceramic wall tiles

CHAPTER 22

External wall finishes

Introduction

The most popular finish to the external wall of a house in the United Kingdom has been facing brickwork. However, in some parts of the country other finishes, such as shiplap boarding, are endemic to particular areas and define the area's vernacular architecture. Timber frame and steel frame construction have been able to use a wide choice of external finishes, since the wall finish does not have to satisfy the loadbearing function that masonry walls are designed to do.

This chapter will consider the functional requirements of an external wall finish and then discuss the application of the three main types of external wall finishes used for housing:

- rendering
- tile hanging
- weatherboarding.

Functional requirements

The main functional requirements that an external wall finish needs to be able to satisfy are:

- weather resistance
- appearance
- durability
- fire resistance.

Weather resistance

To be effective as an external finish the material used must be weather resistant. Indeed some external finishes are placed over brickwork or blockwork walls in order to improve the weather resistance of the wall. External finishes do not, however, need to be impervious.

Appearance

External wall finishes will, to a large extent, define the external appearance of the building. Therefore their appearance will be an important consideration in their selection. Appearance can be influenced by the selection of the material used for the finish and also the textures and colours adopted.

Durability

The external wall finish will be subjected to the vagaries of the climate it is exposed to. Thus careful selection of materials, textures and colours is important. It is likely that the external wall finish will not last the lifetime of the building to which it is attached. Thus at least one replacement in the lifetime of the building should be expected, dependent on materials.

Fire resistance

The external wall finish is unlikely to substantially improve the fire resistance of the material to which it is attached. However, when the material for the external finish has been selected consideration should be given to its surface spread of flame performance.

External wall finishes allow materials such as concrete blockwork to be used for the construction of the external walls of the house. The finish will overcome the main disadvantages of using concrete blockwork on the external face of the wall, which are its poor weather resistance and appearance, but will enable the advantages of concrete blockwork for wall construction to be harnessed, such as the low cost of the material and the speed of laying compared with brickwork.

Rendering

Rendering is a form of external plastering using a mixture of cement, lime, sand and water to provide protection against the penetration of moisture or to provide a desired colour, texture and appearance to an external wall.

For rendering to satisfy the functional requirements of an external wall finish careful consideration needs to be given to the following aspects:

- the mix design
- the bond to the background
- the degree of exposure of the building
- the application of the render finish
- the texture of the surface.

The mix design

Strong cement:sand mixes are impervious but there is a tendency for them to crack during drying or because of differential movement between the finish and the background. Shrinkage cracks allow water to penetrate behind the rendering, which cannot then escape by normal evaporation. This leads to damp walls and loss of adhesion of the render to the backing wall, causing it to eventually fall off.

Weak cement:lime:sand mixes are less likely to crack since there is less shrinkage during drying, but these mixes are more permeable than the stronger mixes. However, moisture entering the render will be able to evaporate as weather conditions change.

The bond to the background

The rendering should not be stronger than the background to which it is attached, otherwise cracking may occur due to differential movement between the finish and the background. Weak mixes are therefore preferable.

A 'spatter dash' coat of 1 part cement to 2 parts sand can be mixed wet and thrown onto the background vigorously to provide a rough surface for the render to bond to. Mortar joints in brickwork and blockwork should be raked out prior to the application of the finish. On concrete surfaces a dovetail key can be provided by using profiled formwork. Alternatively, bitumen coated expanded metal lathing can be attached to the background or to the sheathing on timber frame and steel frame walls to provide a key for the render. Dry backgrounds need wetting to reduce any suction characteristics in the background prior to the application of the render.

On rendered properties, where a gable end wall has been built from solid stone and a brick chimney flue has been built into the wall, the external render can often crack. This originates from the differing thermal coefficients of the brick and the stone in the wall, which causes the two materials to expand at differing rates and places stresses on the render. This vertical cracking can often be misdiagnosed as movement due to subsidence. To prevent this from occurring a strip of expanded metal lathing should be fixed to the wall along the line of the differing materials, before the render is applied. This then allows them to expand and contract at their differing rates without cracking the covering render.

The degree of exposure of the building

Weak mixes are satisfactory in sheltered conditions but not sufficiently durable in exposed conditions. Strong mixes should also be used at points that are vulnerable to impact damage or abrasion.

Weak mixes are slow to harden and therefore may be affected by frost action during the first few weeks after their application. Therefore strong mixes should be used when applying rendering during periods of cold weather.

The application of the render finish

The render may be applied in either two or three coats. Three coats should be used where the background is uneven or where the exposure is severe. Undercoats should be scratched to provide a key for the final coat and allowed to dry thoroughly before the final coat is applied.

The undercoat should be 10mm thick for three coat rendering and 13mm thick for two coat rendering. The second coat in three coat rendering should be 10mm thick. The final coat should be 6mm thick for three coat rendering and 10mm thick for two coat rendering.

The texture of the surface

Five different textures are normally available for rendered finishes:

- smooth
- textured
- rough cast
- pebble dash
- Tyrolean.

Smooth

A steel trowel or wood float finish may be provided. There is a tendency for smooth finishes to *craze* if they are overworked, due to carbonation of the cement *laitance* brought to the surface.

Uneven weathering of the surface, particularly rain staining, can be a problem with this type of finish.

Self-assessment question 22.1

What could cause overworking of a smooth render finish?

Textured

The finish is provided by scraping the final coat with the edge of a trowel, hacksaw blade or nail board. This removes the cement laitance and prevents crazing of the surface.

Rough cast

A wet mix of cement:lime:sand:selected aggregate is thrown onto the undercoat. This provides good adhesion of the final coat to the undercoat. The uneven surface produced by the protrusion of aggregate particles form the final coat of render breaks up water paths and helps to reduce rain staining.

Self-assessment question 22.2

Why should the breaking up of water paths by the protruding aggregate in rough cast rendering be important?

Pebble dash

This is a similar technique to rough cast rendering, but dry aggregate is thrown against the final coat whilst it is still wet and is worked into the render by tapping it with the surface of a trowel. The aggregate used for pebble dash is generally smaller than that used in rough cast finishes. The adhesion of the aggregate to the render is not as good with pebble dash as it is with rough cast.

Self-assessment question 22.3

In what circumstance may a pebble dash finish be preferred to a rough cast finish?

Tyrolean

This is a proprietary composition of render that is flicked onto the undercoat using the blades of a machine to create a decorative textured finish.

When applying rendering to an external wall, it is important to ensure that it does not bridge over the DPC or moisture in the wall below the DPC will be able to travel via the render and enter the wall above the DPC.

The render can be finished just above the DPC by creating a 'bell mouth' detail using an EML edging bead (see Figure 22.1).

If a brick plinth effect is not desired, the render can be re-commenced below the DPC (see Figure 22.2).

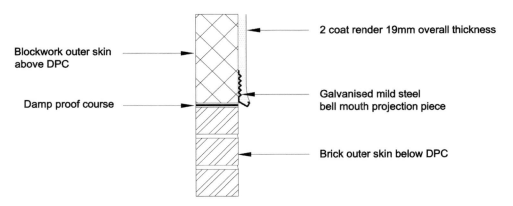

Figure 22.1 Bell mouth detail of rendering at DPC with brick plinth

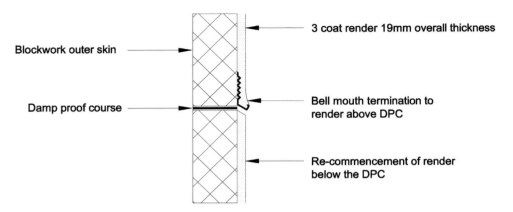

Figure 22.2 Re-commencement of render below the DPC where a brick plinth is not used

Tile hanging

Roof tiles may be used as a finish to an external wall, either to enhance the appearance of the wall or to provide additional weather resistance to the wall. The tiles may be vulnerable to breakage due to impact damage; it may therefore be better to limit their use to more inaccessible areas, such as upper storeys.

Slates and clay or concrete plain tiles can be hung on 38 x 19mm preservative treated battens fixed to the wall. Counter battens may be used in order to ventilate the cavity between the back of the tiles and the wall. These should be fixed at 400mm vertical centres. On timber frame and steel frame houses the battens should be fixed to the sheathing and a breather membrane placed beneath the battens.

Tiles and slates should be double nailed to the battens and laid to a bonded joint as they would be on a roof. Normally a 114mm gauge is used for a standard 265mm tile. Nib-less tiles may be nailed directly to mortar bed joints rather than timber battens, but they must be of a suitable size in order to be able to work in with the brickwork or blockwork courses.

Self-assessment question 22.4

What problem would be encountered if the nib-less tiles were not of a suitable size?

At the head of the wall, tiles should be protected by a cover flashing fixed behind the top batten (see Figure 22.3).

At the foot of the wall, the tiles should be tilted by a tilting fillet (see Figure 22.4).

At openings, tiles may be butted against a projecting frame or specially made corner tiles may be used. At *quoins* corner tiles may be

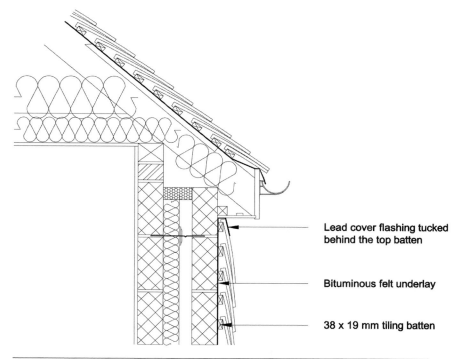

Lead cover flashing tucked
behind the top batten

Bituminous felt underlay

38 x 19 mm tiling batten

Figure 22.3 Cover flashing to tiling at the head of the wall

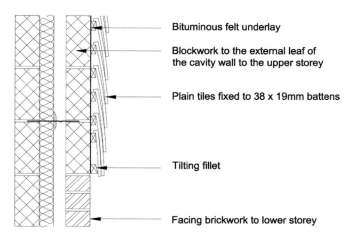

Bituminous felt underlay

Blockwork to the external leaf of
the cavity wall to the upper storey

Plain tiles fixed to 38 x 19mm battens

Tilting fillet

Facing brickwork to lower storey

Figure 22.4 Tilting fillet to tiles at the foot of the wall

used (see Figure 22.5) or tiles may be mitre cut and backed by shaped
lead soakers.

Weatherboarding

Either softwood or hardwood timber boards may be used for
weatherboarding. If non-durable softwoods are used they must be

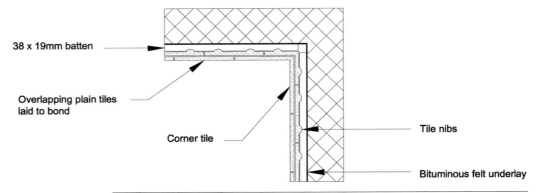

38 x 19mm batten

Overlapping plain tiles
laid to bond

Corner tile

Tile nibs

Bituminous felt underlay

Figure 22.5 Corner tiles

preservative treated by either staining or painting. Western red cedar is very durable and can be very decorative if coated with a clear sealer. Hardwoods may need to be preservative treated but are often left unprotected and allowed to weather naturally. PVCu boards are an acceptable substitute for timber and are very durable but can be prone to impact damage if they are situated in vulnerable areas of the façade.

The boards may be nailed directly to the sheathing of a timber or steel framed house or fixed indirectly to timber grounds plugged to masonry walls. Secret nailing using corrosion resistant metal nails protects the fixings. All boarding needs to be laid on an impervious barrier of bituminous felt or polyethylene sheet, irrespective of the background.

The spacing of the 32 x 25mm preservative treated timber battens to masonry walls is dependent on the thickness of the boards and the manner in which the boards are laid:

- 1200mm spacing for 25mm horizontal boards
- 450mm spacing for 16mm horizontal boards
- 600mm spacing for 25mm vertical boards
- 760mm spacing for 25mm diagonal boards.

The boarding should be free to move on at least one side. The lower edge should be kept at least 150mm above any horizontal surface to prevent splash staining of the boarding. At quoins boards may be crossed and cut or alternatively they may be butted against 75 x 30mm corner boards.

The boards may be fixed in many different designs (see Figure 22.6):

- square edged, horizontal
- feather edged, horizontal
- rebated, feather edged, horizontal
- rebated, shiplap, horizontal
- tongued and grooved with V joint, vertical
- board on board, vertical
- batten on board, vertical.

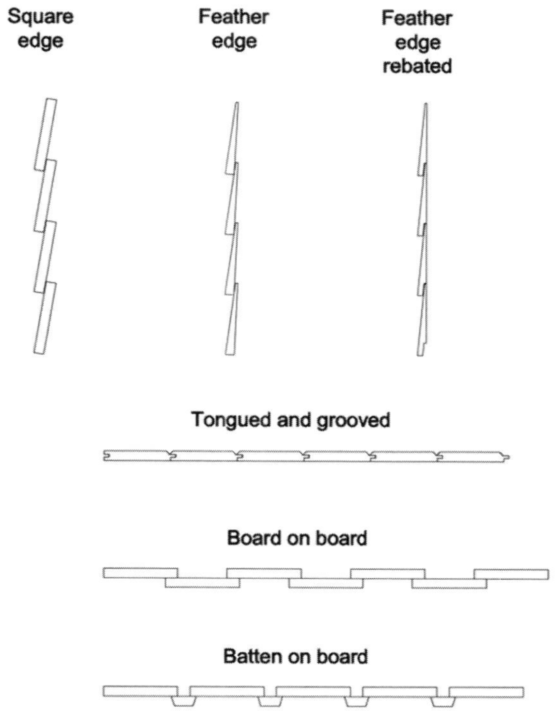

Square
edge

Feather
edge

Feather
edge
rebated

Rebated
shiplap

Tongued and grooved

Board on board

Batten on board

Figure 22.6 Styles of weatherboarding

Visit the companion website to test your
understanding of Chapter 22 with a multiple choice
questionnaire, and to see 26 full-colour photos
illustrating the chapter.

Glossary of terms

Abutment	The point where one part of a building structure adjoins another part of the structure or another building. Sometimes encountered in roof construction
Admixture	A material that can be added to a concrete mix in order to enhance particular characteristics of the concrete such as its workability, frost resistance or setting time
Air entrainment	A method of introducing air or some other gas into concrete before it has achieved its initial set. The bubbles produced reduce the overall weight of the concrete component and provide the advantages of lightweight concrete. Air entrainment can also provide extra resistance to frost in a concrete member
Air permeability	A measure of the amount of air leakage from a building that will affect its thermal performance
Alum	A mineral used in Class C plasters to accelerate their setting time
Angle of repose	The maximum angle at which a soil on a slope is stable
Annealing	A process of toughening glass by heating it and allowing it to slowly cool so that stresses produced in its initial manufacture are relieved
Arch	An arrangement of bricks or stones, usually curved in shape, to carry the wall above over an opening
Architrave	A decorative piece of timber used to cover the joint between the door frame or lining and the wall to which it is attached
Armour plate glass	A toughened glass produced by heating followed by rapid surface cooling producing high compression forces in the surface of the glass
Autoclaving	A method of curing and hardening calcium silicate and concrete bricks and blocks in a steam oven
Balusters	Guarding and safety members usually positioned between the handrail and the outer string of a stairway
Barge board	A board covering the roof construction and covering at the gable end of a pitched roof
Batt	A cut brick. Usually a half brick
Batten	A small timber member often used for fixing other components, particularly roofing tiles
Bearer	A temporary support to a formwork platform or deck. It performs a similar function to a joist in floor construction
Bed face	The bottom face of a brick

Bevelled closer *A brick that is cut longitudinally on the bevel such that one header face is the full width whilst the opposite header face is only half the standard width. It is used when creating stop ends in one and a half brick walls in English or Flemish bond*

Binder *A timber member fixed to the top of ceiling joists at mid-span to prevent excessive deflection*

Bird's mouth joint *The joint between the rafter and the wall plate*

Bottom plates or sill plates *The bottom member in a timber framework*

Braces *Diagonal members fixed between vertical and horizontal frame members to help keep the frame square and prevent racking*

Breather membrane *A material that is moisture vapour permeable but water liquid impermeable. When fixed to the sheathing in a timber frame wall any excess water vapour trapped in the wall may escape to the outside air, whereas any rain that may have penetrated the external wall finish will be unable to cross the breather membrane and cause dampness in the timber construction of the wall*

Browning *An undercoat plaster generally used on masonry*

Building Line *A nominal line used by Town and Country Planning Authorities to determine the position where building may commence relative to the front boundary of the property*

Bulbs of pressure *The lines drawn between points of equal pressure in the ground when the soil is loaded by a building foundation*

Calendering *A process in which a sheet is passed between rollers (often heated) to smooth it or to bond the surface finish to the background*

Cambium *A soft substance immediately beneath the bark of a tree in which the annual growth of the wood and bark takes place*

Carbohydrates *Organic compounds of carbon, oxygen and hydrogen often forming sugars that the tree will feed upon*

Carcassing *Timber forming the framework to a building or parts of a building. Carcassing timber is rarely seen after the completion of the building*

Casement *A frame in a window that can be opened*

Cellulose *A substance that forms the essential part to the solid framework of wood*

Centre *Temporary support when constructing an arch*

Charring *The production of charcoal by the partial burning of wood*

Chimney *A structure, normally in brickwork, that protrudes through the roof and provides a means for the products of combustion from open fires or boilers to be vented to the atmosphere*

Cill or sill *The bottom of an opening (usually a window opening; see threshold for doors)*

Close couple roof *A roof similar to a couple roof but having a ceiling joist to tie the rafters together at their base*

Collar roof *A roof similar to a close couple roof but having the ceiling joists raised up the roof to provide extra height to the upper storey*

Combed joint	*A joint used in door and window frame construction to connect members together. It comprises a number of interconnecting 'teeth'*
Compartmentation	*The division of a building into a number of fire resisting compartments to reduce the spread of fire in the building and also to provide a safe means of escape from the building in the event of a fire*
Coniferous	*A tree that bears cones and normally retains its foliage throughout the year*
Coniophora Puteana, formerly known as Coniophora Cerebella	*A species of fungi that is commonly the cause of wet rot in timber*
Coping	*The capping to a parapet*
Counter batten	*A length of timber of similar size to a tiling batten but running up the slope of the roof at similar spacing to that of the rafters, to enable the tiling battens to be raised above the surface of the sarking felt where roof boarding has been placed over the rafters*
Couple roof	*A simple pitched roof comprising rafters bearing on wall plates at their feet and a ridge board at their heads*
Cover	*The thickness of concrete in a structural member that protects the steel reinforcement from the effects of fire and corrosion*
Crazing	*The formation of very fine cracks in the surface of glazed ceramic tiles or plaster*
Curing	*The process of setting and hardening in concrete*
Dado rail	*The capping used on timber panelling that is terminated approximately one metre above the floor level*
Deciduous	*A tree that sheds its leaves each year*
Dew point temperature	*The temperature at which a quantity of air becomes saturated by a quantity of water vapour*
Dimensional co-ordination	*The standardisation of dimensions for components in a building that is agreed amongst the manufacturers, allowing similar components from differing manufacturers to be interchanged for incorporation into a building*
Doorset	*An internal door supplied already hung in its frame*
Double lap tiling	*Roof tiles that require an overlap at their heads and bonding at their sides*
Double roof	*A roof having the rafters supported at mid-span by purlins*
Dragline	*A machine that can excavate large quantities of soil at large distances or depths from the machine*
Dry lining	*An internal finish that does not use wet materials. Normally plasterboard is used*
Dry rot	*Species of fungi that feed on damp rather than wet timber*
Dual pitch	*A roof pitched or sloping on two sides*
Ductility	*The property of a material that can be bent without cracking*
Eaves	*The edge of a roof where a rainwater gutter is normally situated*
Edge trim	*A strip of metal, usually aluminium, fixed to the edge of a flat roof and into which the mastic asphalt or bituminous felt roof covering is dressed*

Efflorescence   *A white powder that can appear on the surface of clay bricks. It is caused by soluble salts being brought to the surface of the bricks by moisture as it evaporates*

Elasticity   *An engineering term used for materials that will deform under load but will return to their original shape when the load is removed*

Emissivity   *The radiating power of heat from a material such as glass in a window*

Factor of safety   *A figure used in structural calculations to increase the anticipated building load so that the designed structural component is capable of safely carrying loads in excess of the anticipated load*

Fascia board   *A board, normally of timber, that covers the ends of the joists or rafters in a roof at the eaves or verge*

Fenestration   *The architectural styling and disposition of windows within the façade of a building*

Fines   *Aggregate that has a particle size of less than 5mm*

Flanges   *The top and bottom horizontal sections to an I section beam*

Flashing   *A strip of malleable and impervious material, usually sheet metal such as lead or copper, used to cover and weatherproof the gap between the roof covering and a projecting feature such as a chimney stack*

Flight   *A series of continuous steps that are not interrupted by a landing*

Flintlimes   *Bricks that are made from a combination of flint aggregates and lime*

Fluxes   *A substance used to aid the fusion of ceramic materials by heating them during manufacture*

Folding wedges   *Wedge-shaped pieces of timber that provide fine adjustment and tightening of props or trench struts to ensure that the load is safely transferred*

Formwork or shuttering   *A temporary framework or platform, normally constructed of timber, that supports the wet concrete in its final shape whilst it is gaining sufficient strength to be self-supporting*

Frass   *The refuse left behind by the larvae of boring insects*

Frog   *An indentation in the top face of a brick that reduces its weight and speeds up the hardening process in the kiln*

Furrings   *Galvanised mild steel channel sections that can be used in preference to timber battens for fixing plasterboard dry lining to walls*

Gable end   *The end of a building that is not covered by the roof structure. The gable wall is built up to the underside of the roof covering*

Gable ladder   *A support to the barge board at the gable end of a pitched roof*

Gauge   *The spacing between tiling battens. The gauge also determines the head lap of the roof tiles*

Glazier's sprigs   *Small tacks that hold the glass into the timber casement rebate prior to the application of putty*

Going   *The useable part of the tread. Generally the distance between nosings*

Grading   *The arrangement of soils or aggregates by particle size*

Granolithic   *A hardwearing floor finish containing a granite aggregate. Usually applied as a monolithic screed*

Grounds	Timber battens used to provide a straight and level surface for the fixing of wall strings, skirtings or rails used in wall panelling
Grout	The fine mixture of water, cement particles and very fine aggregate particles that form on the surface of concrete when it has been placed and compacted
Gypsum	A mineral that consists of a crystalline combination of calcium sulphate and water. Used in the manufacture of plasters and some cements
Half space landing	A horizontal platform between flights that turns the stairway through 180^0
Handrail	A protecting and safety member usually positioned on the outside of a stairway. It can also assist people who need support when ascending and descending the stairs
Hanger	A timber member fixed to the binder at its foot and the ridge board at its head
Head	The horizontal top member in a door or window frame
Head plate	The top member in a timber framework
Header face	The end face of a brick
Headroom	The vertical distance between the pitch line of the flight of a stairway and the underside of the floor, landing or flight above
Herringbone strutting	40 x 40mm diagonal braces placed between joists in upper floors to act as horizontal restraint to prevent twisting of the joists in their length due to moisture movement
Hipped roof	A roof pitched or sloping on three or four sides
Humidity	A measure of the amount of water vapour held in suspension by the air
Hydration	The process of cement setting and hardening
Hygroscopic	Materials that are sensitive to moisture and which can raise or lower their moisture content relative to the prevailing humidity level of the surrounding environment
Hyphae	White strands emanating from the spore of a fungus that spread across affected timber and feed on its cellulose
Inclined struts	Support to purlins to prevent excessive deflection
Integrity	A rating that is related to the ability of a door to resist the passage of flame for the designated period of fire resistance for the category of the door
Interstitial condensation	The condensing of water vapour into water droplets within the structure of the building
Intumescent strip	A material that can be added to the edge of a fire door or its associated frame and which swells when subjected to a high temperature. This seals the gap between the door and the frame thus preventing the passage of smoke, flames and hot gases
Isolating membrane	A material generally used in flat roof construction to separate the mastic asphalt or sheet metal roof covering from the timber deck or the concrete slab of the roof construction. This allows differential movement between the covering and the roof structure to occur without causing stresses in either material

Jamb	The vertical side member in a door or window frame
Just in time	A system of delivery management in which the timing of material and component deliveries to site is planned to coincide with their incorporation into the building, so reducing the amount of storage required on site
Jute	A plant fibre that can be woven into a material and used as the backing for linoleum
Jute scrim	A strip of fabric used to reinforce the joints between sheets of baseboard plasterboard and prevent cracking in the skim coat of plaster applied as a finish to the surface of the plasterboard
Kentledge	Heavy static weights that are placed on top of a pile to test its loadbearing capacity
Keratin	An animal protein incorporated into Class B plasters to slow down their setting time
Keystone	The brick or stone at the centre of an arch
Laitance	A fatty substance present in cement grout that can be brought to the surface when cement-based renders are overworked. Its presence on the surface of smooth renders can lead to the surface developing very fine cracks or crazing
Laminas	Thin sections of timber that can be glued together to form a larger section. Often used in the manufacture of laminated timber beams or laminboard
Landing	The horizontal platform between flights in a stairway
Latex	Unrefined rubber
Lean to roof	A monopitch roof in which the rafters span between an abutment wall at their heads and a supporting wall at their feet
Ledges	Horizontal timbers used in matchboarded door construction to clamp the battens of the door panel together
Life cycle costing	The total cost of a component over its entire life cycle including projected maintenance and replacement costs as well as the initial purchase and installation cost of the product
Lignification	The process of producing lignin
Lignin	An organic substance forming the essential part of wood fibre
Lintel or lintol	A concrete or steel beam that carries the wall above over an opening
Lipping	A thin strip of timber fixed at the edges of a door to protect them from damage and to provide a tolerance for planing the doors to fit the frame or door lining when the doors are hung into position
Mansard roof	A roof where the lower portion is pitched at a steeper angle than the upper portion
Modular co-ordination	The standardisation of incremental dimensions for components in a building that is agreed amongst the manufacturers
Module	A standard unit of measurement used in the sizing of components to provide incremental variations in size within the range

Modulus of elasticity	*The ratio of stress to strain that provides a measure of the elasticity of a structural material*
Mohr's circles	*The plot of principle stress against shear stress of a soil sample tested in a tri-axial test apparatus. This provides a measure of the shear strength of the soil sample being tested*
Monopitch	*A roof pitched or sloping on one side only*
Mortar	*A mixture of cement, sand and usually lime with water that provides the bedding for bricks, blocks or stones in the building of a wall*
Mortice and tenon joint	*A joint between two timber components. The tenon is a thin projection from one of the members being joined that fits into a sinking of similar size (the mortice) in the other timber member being joined*
Mullion	*A vertical sub-dividing member in a window frame*
Mycelium	*A mass of hyphae resembling in appearance a ball of cotton wool*
Nail plate connectors	*A galvanised metal plate used to connect timber members together in a prefabricated trussed rafter*
Neoprene gasket	*A synthetic rubber material that is used to fit glazing into aluminium and PVCu window and door frames, usually in the factory*
Newel post	*A member in a staircase that houses the outer string and supports it*
Node points	*The points at which members in a trussed rafter are connected together*
Nogging	*A short piece of timber fixed between vertical timber studs in an internal partition wall or between joists in an upper floor in order to stiffen the construction*
Open riser stair	*A staircase that does not use risers to form the steps*
Parapet	*The part of a wall that protrudes above the roof*
Parenchyma	*Brick-shaped food storage cells found in softwoods and hardwoods*
Party wall	*A shared wall between properties in semi-detached or terrace housing*
Perpend	*The vertical joint between adjacent bricks*
Photosynthesis	*The conversion by plants of carbon dioxide and water into carbohydrates brought about by exposure to sunlight*
Piles	*A concrete foundation that extends into the ground and can carry the load of a building to a suitable bearing stratum. Also the loops in a carpet forming the nap or surface finish*
Pitch	*The angle of inclination of the stairs*
Pitch line	*A notional line connecting the nosings of all the treads in the flight of a stairway*
Planking and strutting	*The method of supporting the sides of an excavation. Traditionally timber has been used for this purpose*
Planted stop	*A piece of timber that is fixed to the jamb of a door frame or door lining and onto which the door closes*
Plaster of Paris	*A quick setting plaster. Useful for repair work*
Plasticisers	*An admixture that can be added to mortar or cement mixes to improve their workability*

Plies	*Veneers of timber that can be glued together to form a strong and dimensionally stable board*
Poling boards	*Vertical boards placed against the sides of an excavation to support them*
Profile boards	*Boards to indicate the position of trenches on a site*
Protoplasm	*A semi-fluid and semi-transparent substance consisting of oxygen, hydrogen, carbon and nitrogen, along with traces of other elements, providing the physical basis of life in all plants and animals*
Pug mill	*A machine that mixes in water with excavated clay to increase the plasticity of the clay so that it may be moulded into bricks*
Purlin	*A timber member supporting the rafters at mid-span to prevent excessive deflection*
Quarter space landing	*A horizontal platform between flights that turns the stairway through 90^0*
Quoin	*An external corner of a building*
Racking	*The dropping out of square of a frame*
Rafter	*The principal loadbearing member in a pitched roof. It performs a similar function to that of a joist in floor or flat roof construction*
Reinforced concrete	*Concrete that contains a material (usually steel) that has high tensile strength and which can work in combination with the concrete to produce a structural material that has good resistance to compressive, tensile and shear forces*
Render	*A mixture of cement and sand and often lime that is applied to the external surface of a wall*
Rhizomorphs	*Thin strands emanating from the mycelium and conveying moisture to previously unaffected timber, making it suitable for attack by the hyphae*
Ridge	*The top of a pitched roof*
Ridge board	*A timber member providing a fixing for the rafters at their heads*
Rise	*The vertical distance between two consecutive treads in a stairway*
Riser	*The vertical component of a step*
Rising butt hinges	*Hinges that have a spiral knuckle allowing the door to rise as it opens so that it can clear the floor finishes easily. Because the flap of this hinge that is fixed to the door can be lifted off the central pin holding the two flaps of the hinge together, this hinge is often used on doorsets enabling the door to be removed from its frame prior to installation*
Rolled steel joist (RSJ)	*A small structural steel beam used to carry relatively heavy loads over a small span*
Rough bracket	*A piece of timber that is fixed to the underside of a tread in a staircase to provide extra support, particularly at mid-span. It is fixed to a bearer that runs the length of the flight beneath the staircase*
Runners	*Boards placed against the sides of an excavation to support them. Their ends are driven into the ground below the excavation to provide further support*
Sacrification	*The process of over-sizing structural timber members so that, should a fire occur, the extra sacrificial timber will be consumed by the fire and this will not affect the functioning of the structural timber member*

Sandlimes	Bricks that are made from a combination of sand and lime
Sap	A fluid mainly consisting of water and containing essential minerals to feed the tree
Sarking membrane	A secondary weatherproof barrier on pitched roofs. This may be bituminous felt reinforced with hessian or reinforced polyethylene sheet
Scarf joint	A joint used to join rafters or purlins together
Scribing	Cutting one member to fit the profile of another where they join
Seasoning	A technique for reducing the moisture content of recently converted timber to levels that are appropriate for its end use
Segregation	The separation of coarse and fine aggregate particles that may occur if concrete is poured from a great height when being placed. The resulting concrete is not the homogeneous mix that is desired and its properties may be impaired
Serpula Lacrymans, formerly known as Merulius Lacrymans	A species of fungi that is commonly the cause of dry rot in timber
Sheathing	Plywood sheeting fixed to the exterior of a timber frame panel
Sheeting boards	Horizontal boards placed against the sides of an excavation to support them
Single lap tiling	Roof tiles that require an overlap at their heads but use an overlapping joint at their side or a patent edge fixing to adjacent tiles so that they do not need to be bonded at the sides
Sintered pulverised fuel ash	A by-product from coal fired power stations. It has similar setting characteristics to cement and can be added to some concrete mixes to reduce the amount of cement used
Sintering	The fusing together of particles of material, usually by heat
Skew nailing	Nailing at an angle and not perpendicular to the timber member being fixed
Skirting board	A decorative piece of timber used to cover the joint between the wall and the floor
Slenderness ratio	The ratio between the effective height and the effective thickness of a structural member in compression (generally a wall or column). A high slenderness ratio can lead to buckling in the member when it is loaded
Soffit	The underside of a structure. In floors and roofs this is normally referred to as the ceiling
Soffit board	A board, normally of timber, that covers the underside of exposed joists or rafters in a roof at the eaves or verge
Sole piece	A large piece of timber that supports the raking strut at its base
Sole plate or base plate	A timber member usually positioned at ground level, onto which timber frame panels are fixed
Spalling	The breaking up of the surface of a brick. Generally caused by the action of frost on clay bricks
Split ring connector	A galvanised metal ring used to connect timber members together in a prefabricated truss. The ring is split to enable the members to move relative to each other but still remain connected

Split spoon sampler	*An apparatus used to establish the relative density of a soil in situ*
Sporophore	*The fruiting body of a fungus*
Sprocket	*A timber member that is used to support the underlay in a pitched roof and also enables the last course of tiles to be raised so that they can drain into the rainwater gutter*
Stairwell	*The space that a stairway occupies in a building*
Stretcher face	*The side face of a brick*
String	*A member in a staircase that houses the treads and risers in a stairway and supports them*
Struts	*A supporting member placed across an excavation to support the poling boards or sheeting boards and prevent them from collapsing inwards by the pressure of the earth in the sides of the excavation. The member is in compression and the term may also be applied to other members in a structure that are subject to compression forces, particularly in a roof*
Stucco	*A mixture of lime, sand and linseed oil that was used as an external finish to walls in the eighteenth and nineteenth centuries*
Stud	*A vertical timber member used in a timber framework*
Stud partitions	*A framework of timber or light galvanised steel members with plasterboard facings to produce a lightweight internal wall that is usually non-loadbearing*
Surface condensation	*The condensing of water vapour into water droplets on cold surfaces of the building*
Tapered treads or winders	*Treads that turn the stairway through an angle. They are narrower on one side than the other*
Terrazzo	*A decorative floor finish containing a marble aggregate. Usually applied as a monolithic screed*
Thermal bridging	*The introduction of a material having a high thermal conductivity value which by-passes the thermal insulation material in a building element causing heat to escape from a building*
Thermal conductivity	*The measure used to determine the ability of a material to conduct heat through it*
Thermal insulant	*A material with a low thermal conductivity which does not readily conduct heat through it*
Thermal mass	*The ability of a building to retain heat*
Thermal resistance	*A measure of the thermal resistivity of a material based on its thickness*
Thermal resistivity	*The measure used to determine the resistance of a material to conduct heat through it. The inverse of thermal conductivity*
Threshold	*The bottom member of a door frame, similar to the cill in a window frame*
Throating	*A groove on the underside of a cill member or the back of a jamb member to prevent water from travelling across the member by capillary action*
Tie	*A structural member in tension*
Toothed plate connector	*A galvanised metal plate used to connect timber members together in a prefabricated truss*

Torque	*The twisting or rotating force applied to an object*
Tracheid	*A pod like cell used for the conduction of sap in wood*
Transom	*A horizontal sub-dividing member in a window frame*
Traveller	*A rod of a pre-determined length that, when used with profile boards, enables the depth of a trench to be ascertained*
Tread	*The horizontal component of a step*
Tricalcium aluminate	*A chemical component of cement that can make concrete susceptible to attack by soluble sulphates if the amount present in the cement is too great*
Triple roof	*A roof having the purlins supported by a prefabricated truss at every fourth set of rafters*
Truss plates	*A galvanised metal plate that fixes a trussed rafter to the wall plate*
Trussed rafter	*The arrangement of timber sections in a factory into a structurally stable roof truss*
Trussed rafter roof	*Self-supporting prefabricated trusses that span between wall plates and dispense with purlins, ridge board, hangers and binders*
U-value	*The air to air thermal transmittance coefficient of an element made up of the sum of the thermal resistances of the component materials and surfaces*
Undercloak	*A board used at the verge to a pitched roof to support the mortar filling to the underside of the roof tiles*
Underlay	*A sheet material, usually of rubber or felt, used as a base layer for carpets that do not have a latex backing*
Valley roof	*A roof where two sloping surfaces meet at an internal intersection*
Vapour check	*A layer of generally impermeable material that can be used to reduce the passage of water vapour across it. It cannot be considered to be 100 per cent impervious to the passage of water vapour as it may be punctured by nails or staples used to fix it or other materials to the structure*
Vapour control layer	*A layer of impervious material that will prevent the passage of water vapour across it. It can be 100 per cent efficient as it will be continuous and unbroken*
Veneer	*A thin slice of timber often used to provide a decorative effect to wood products*
Ventlight	*A top hung window casement providing a small amount of ventilation when required*
Verge	*The edge of a roof where a rainwater gutter is not situated*
Vitrification	*The creation of a glass-like substance by heating certain ceramic materials*
Voussoirs	*Wedged-shaped bricks used in the construction of gauged arches*
Vulcanising	*The treatment of natural rubber with sulphur and subjecting it to intense heat to improve its strength, wear resistance and durability*
Wainscoting	*Timber panelling generally up to a height of approximately one metre above the floor level*

Walings   *A member, usually timber, that is placed across the face of a series of poling boards or sheeting boards in order to spread the support from the struts*

Wall plate   *A timber member providing a fixing for rafters or joists on the supporting wall*

Warping   *The twisting or bending out of shape of timber*

Water bar   *A galvanised mild steel bar incorporated into a cill or threshold to prevent rainwater from running across the top surface of the member and entering the building. Also a rubber section used in waterproof concrete construction to prevent water entering a basement through the joint between concrete wall or floor sections*

Weatherboard   *A sloping member fixed to the bottom member of a door to throw water clear that has run down the face of the door*

Weatherstripping   *The addition of a thin layer of compressible material to the edges of an opening window casement or door to improve its draught exclusion capability. It can also reduce the amount of heat lost and sound transmitted through the gap between the window casement or door and its associated frame*

Web   *The middle vertical section of an I section beam*

Weepholes   *Apertures formed in the vertical mortar joints of a cavity wall to enable trapped moisture to be drained to the outside*

Wet rot   *Species of fungi that feed on wet timber. Often also referred to as cellar rot*

Window board   *A timber or plastic board that covers the inner part of the wall at the cill and is joined to the back of the bottom section of the window frame*

Workability   *The ability of a concrete mix to be mixed, transported, placed and compacted effectively so that the concrete may achieve its intended purpose*

Yarn   *Fibre spun for use in the weaving of carpets*

Bibliography

Accredited Construction Details, Department for Communities and Local Government, 2007

Assessing moisture in building materials – Part 1, Sources of moisture, Good Repair Guide GR33/1, Building Research Establishment, 2002

Assessing moisture in building materials – Part 2, Measuring moisture content, Good Repair Guide GR33/2, Building Research Establishment, 2002

Assessing moisture in building materials – Part 3, Interpreting moisture data, Good Repair Guide GR33/3, Building Research Establishment, 2002

Best practice guide to timber fire doors: manufacture, specification, installation, approval and maintenance, Architectural and Specialist Door Manufacturers Association, 2002

BRE Green Guide to Specification, 4th edn, Building Research Establishment, 2009

Bricks, blocks and masonry made from aggregate concrete – Part 1 Performance requirements, Digest DG460/1, Building Research Establishment, 2001

Bricks, blocks and masonry made from aggregate concrete – Part 2 Appearance and environmental aspects, Digest DG460/2, Building Research Establishment, 2001

British Wood Preserving and Damp Proofing Association (BWPDA) Manual (2000 revision), British Wood Preserving and Damp Proofing Association, 2000

BS 146:2002 Specification for blastfurnace cements with strength properties outside the scope of BS EN 197-1

BS 459:1988 Specification for matchboarded wooden doors for external use

BS 476-10:2009 Fire tests on building materials and structures. Guide to the principles, selection, role and application of fire testing and their outputs

BS 476-22:1987 Fire tests on building materials and structures. Methods for determination of the fire resistance of non-loadbearing elements of construction

BS 476-31:1983 Fire tests on building materials and structures. Methods for measuring smoke penetration through doorsets and shutter assemblies. Method of measurement under ambient temperature conditions

BS 585-1:1989 Wood stairs. Specification for stairs with closed risers for domestic use, including straight and winder flights and quarter or half landings

BS 644:2009 Wood windows. Fully finished factory-assembled windows of various types. Specification

BS 743:1970 Specification for materials for damp proof courses

BS 1187:1959 Specification for wood blocks for floors

BS 1243:1978 Specification for metal ties for cavity wall construction

BS 1370:1979 Specification for low heat Portland cement

BS 1711:1975 Specification for solid rubber flooring

BS 3444:1972 Specification for blockboard and laminboard

BS 4027:1996 Specification for sulphate resisting Portland cement

BS 4787-1:1980 Internal and external wood doorsets, door leaves and frames. Specification for dimensional requirements

BS 4873:2009 Aluminium alloy windows and doorsets. Specification

BS 4978:2007 + Amendment 1:2011 Visual strength grading of softwood. Specification

BS 5250:2011 Code of practice for control of condensation in buildings

BS 5268-2:2002 Structural use of timber. Code of practice for permissible stress design, materials and workmanship

BS 5268-3:1998 Structural use of timber. Code of practice for trussed rafter roofs

BS 5268-6.1:1996 Structural use of timber. Code of practice for timber frame walls. Dwellings not exceeding seven storeys

BS 5270-1:1989 Bonding agents for use with gypsum plasters and cement. Specification for polyvinyl acetate (PVAC) emulsion bonding agents for internal use with gypsum building plasters

BS 5395-1:2010 Stairs. Code of practice for the design of stairs with straight flights and winders

BS 5588-1:1990 Fire precautions in the design, construction and use of buildings. Code of practice for residential buildings

BS 5628-1:1992 Code of practice for use of masonry. Structural use of unreinforced masonry

BS 5628-3:2001 Code of practice for use of masonry. Materials and components, design and workmanship

BS 5930:1999 + Amendment 2: 2010 Code of practice for site investigation

BS 5950-5:1998 Code of practice for design of cold formed thin gauge sections

BS 6398:1983 Specification for bitumen damp proof courses for masonry

BS 6399-1:1996 Loading for buildings. Code of practice for dead and imposed loads

BS 6399-2:1997 Loading for buildings. Code of practice for wind loads

BS 6399-3:1998 Loading for buildings. Code of practice for imposed roof loads

BS 6510:2010 Steel framed windows and glazed doors. Specification

BS 6515:1984 Specification for polyethylene damp proof courses for masonry

BS 6826:1987 Specification for linoleum cork carpet sheet and tiles

BS 6925:1988 Specification for mastic asphalt for building and civil engineering (limestone aggregate)

BS 7412:2007 Specification for windows and doorsets made from unplasticized polyvinylchloride (PVC-U) extruded hollow profiles

BS 7543:2003 Guide to durability of buildings and building elements, products and components

BS 8004:1986 Code of practice for foundations

BS 8103-1:1995 Structural design of low-rise buildings. Code of practice for stability, site investigation, foundations and ground floor slabs for housing

BS 8103-2:1996 Structural design of low-rise buildings. Code of practice for masonry walls for housing

BS 8103-3:1996 Structural design of low-rise buildings. Code of practice for timber floors and roofs for housing

BS 8103-4:1995 Structural design of low-rise buildings. Code of practice for suspended concrete floors for housing

BS 8110-1:1997 Structural use of concrete. Code of practice for design and construction

BS 8214:2008 Code of practice for fire door assemblies
BS 8215:1991 Code of practice for design and installation of damp proof courses in masonry construction
BS 8218:1998 Code of practice for mastic asphalt roofing
BS 8417:2011 Preservation of wood. Code of practice
BS 8500-1:2006 Concrete. Method of specifying and guidance for the specifier
BS 8747:2007 Reinforced bitumen membranes for roofing. Guide to selection and specification
BS CP 102:1973 Code of practice for protection of buildings against water from the ground
BS EN 197-1:2000 Cement. Composition, specifications and conformity criteria for common cements
BS EN 197-4:2004 Cement. Composition, specifications and conformity criteria for low early strength blastfurnace cements
BS EN 206-1:2000 Concrete. Specification, performance, production and conformity
BS EN 300:2006 Oriented strand board (OSB). Definitions, classification and specifications
BS EN 312:2003 Particleboards. Specifications
BS EN 314-2:1993 Plywood. Bonding quality. Requirements
BS EN 335-1:2006 Durability of wood and wood-based products. Definitions of use classes. General
BS EN 335-2:2006 Durability of wood and wood-based products. Definitions of use classes. Application to solid wood
BS EN 335-3:1996 Hazard classes of wood and wood-based products against biological attack. Application to wood-based panels
BS EN 338:2009 Structural timber. Strength classes
BS EN 350-1:1994 Durability of wood and wood-based products. Natural durability of solid wood. Guide to the principles of testing and classification of natural durability of wood
BS EN 350-2:1994 Durability of wood and wood-based products. Natural durability of solid wood. Guide to natural durability and treatability of selected wood species of importance in Europe
BS EN 351-1:2007 Durability of wood and wood-based products. Natural durability of solid wood. Preservative treated solid wood. Classification of preservative penetration and retention
BS EN 460:1994 Durability of wood and wood-based products. Natural durability of solid wood. Guide to the durability requirements for wood to be used in hazard classes
BS EN 490:2011 Concrete roofing tiles and fittings for roof covering and wall cladding. Product specifications
BS EN 520:2004 + Amendment 1:2009 Gypsum plasterboards. Definitions, requirements and test methods
BS EN 594:1996 Timber frame structures. Racking strength and stiffness of timber frame wall panels
BS EN 599-1:2009 Durability of wood and wood-based products. Efficacy of preventive wood preservatives as determined by biological tests. Specification according to use class
BS EN 599-2:1997 Durability of wood and wood-based products. Performance of preservatives as determined by biological tests. Classification and labelling
BS EN 622-1:2003 Fibreboards. Specifications. General requirements
BS EN 622-2:2004 Fibreboard. Specification. Requirements for hardboards

BS EN 622-4:2009 Fibreboard. Specification. Requirements for softboards

BS EN 622-5:2009 Fibreboard. Specification. Requirements for dry process boards (MDF)

BS EN 634-2:2007 Cement-bonded particleboards. Specifications. Requirements for OPC bonded particleboards for use in dry, humid and external conditions

BS EN 635-1:1995 Plywood. Classification by surface appearance. General

BS EN 635-2:1995 Plywood. Classification by surface appearance. Hardwood

BS EN 635-3:1995 Plywood. Classification by surface appearance. Softwood

BS EN 636: 2003 Plywood. Specifications

BS EN 649:2011 Resilient floor coverings. Homogenous and hetrogenous polyvinylchloride floor coverings. Specification

BS EN 771-1:2003 Specification for masonry units. Clay masonry units

BS EN 771-2:2011 Specification for masonry units. Calcium silicate masonry units

BS EN 771-3:2011 Specification for masonry units. Aggregate concrete masonry units (dense and lightweight aggregates)

BS EN 772-1:2011 Methods of test for masonry units. Determination of compressive strength

BS EN 772-5:2001 Methods of test for masonry units. Determination of the active soluble salts content of masonry units

BS EN 772-6:2001 Methods of test for masonry units. Determination of bending tensile strength of aggregate concrete masonry units

BS EN 772-7:1998 Methods of test for masonry units. Determination of water absorption of clay masonry damp proof course units by boiling in water

BS EN 772-10:1999 Methods of test for masonry units. Determination of moisture content of calcium silicate and autoclaved aerated concrete units

BS EN 772-11:2011 Methods of test for masonry units. Determination of water absorption of aggregate concrete, autoclaved aerated concrete, manufactured stone and natural stone masonry units due to capillary action and the initial rate of water absorption of clay masonry units

BS EN 772-14:2002 Methods of test for masonry units. Determination of moisture movement of aggregate concrete and manufactured stone masonry units

BS EN 772-16:2011 Methods of test for masonry units. Determination of dimensions

BS EN 772-21:2011 Methods of test for masonry units. Determination of water absorption of clay and calcium silicate masonry units by cold water absorption

BS EN 845-1:2003+Amendment 1:2008 Specification for ancillary components for masonry. Ties, tension straps, hangers and brackets

BS EN 942:2007 Timber in joinery. General requirements

BS EN 1172:2011 Copper and copper alloys. Sheet and strip for building purposes

BS EN 1303:2005 Building hardware. Cylinders for locks. Requirements and test methods

BS EN 1304:2005 Clay roofing tiles and fittings. Product definitions and specifications

BS EN 1634-1:2008 Fire resistance and smoke control tests for door, shutter and openable window assemblies and elements of building hardware. Fire resistance tests for doors, shutters and openable windows

BS EN 1634-3:2004 Fire resistance and smoke control tests for door and shutter assemblies, openable windows and elements of building hardware. Smoke control test for door and shutter assemblies

BS EN 1906:2010 Building hardware. Lever handles and knob furniture. Requirements and test methods

BS EN 12051:2000 Building hardware. Door and window bolts. Requirements and test methods

BS EN 12104:2000 Resilient floor coverings. Cork floor tiles. Specification

BS EN 12326-1:2004 Slate and stone products for discontinuous roofing and cladding. Product specification

BS EN 12588:2006 Lead and lead alloys. Rolled lead sheet for building purposes

BS EN 12620:2002 + Amendment 1:2008 Aggregates for concrete

BS EN 13055-1:2002 Lightweight aggregate. Lightweight aggregates for concrete, mortar and grout

BS EN 13162:2008 Thermal insulation products for buildings. Factory made mineral wool products. Specification

BS EN 13163:2008 TIPFB. Factory made products of expanded polystyrene. Specification

BS EN 13164:2008 TIPFB. Factory made extruded polystyrene foam products. Specification

BS EN 13165:2008 TIPFB. Factory made rigid polyurethane foam products. Specification

BS EN 13166:2008 TIPFB. Factory made products of phenolic foam. Specification

BS EN 13167:2008 TIPFB. Factory made cellular glass products. Specification

BS EN 13279-1:2008 Gypsum binders and gypsum plasters. Definitions and requirements

BS EN 13297:2000 Textile floor coverings. Classification of needle pile floor coverings

BS EN 13658-1:2005 Metal lath and beads. Definitions, requirements and test methods. Internal plastering

BS EN 13707 + Amendment 2:2009 Flexible sheets for waterproofing. Reinforced bitumen sheets for roof waterproofing. Definitions and characteristics

BS EN 13914-2: 2005 Design, preparation and application of external rendering and internal plastering. Design considerations and essential principles for internal plastering

BS EN 13967:2004 + Amendment 1:2006 Flexible sheets for waterproofing. Plastic and rubber damp proof sheets including plastic and rubber basement tanking sheets. Definitions and characteristics

BS EN 13969:2004 Flexible sheets for waterproofing. Bitumen damp proof sheets including bitumen basement tanking sheets. Definitions and characteristics

BS EN 14076:2004 Timber stairs. Terminology

BS EN 14081-1:2005 + Amendment 1:2011 Timber structures. Strength graded structural timber with rectangular cross section. General requirements

BS EN 14411:2006 Ceramic tiles. Definitions, classification, characteristics and marking

BS EN 14499:2004 Textile floor coverings. Minimum requirements for carpet underlays

BS EN 14647:2005 Calcium aluminate cement. Composition, specifications and conformity criteria

BS EN 15197:2007 Wood-based panels. Flaxboards. Specifications

BS EN 15217:2007 Energy performance of buildings. Methods of expressing energy performance and for energy certification of buildings

BS EN 15643-2:2011 Sustainability of construction works. Assessment of buildings. Framework for the assessment of environmental performance

BS EN 15644:2008 Traditionally designed prefabricated stairs made of solid wood. Specifications and requirements

BS EN 15743:2010 Supersulphated cement. Composition, specifications and conformity criteria

BS EN 21930:2007 Sustainability in building construction. Environmental declaration of building products

BS EN ISO 6946:2007 Building components and building elements. Thermal resistance and thermal transmittance. Calculation method

BS EN ISO 13370:2007 Thermal performance of buildings. Heat transfer via the ground. Calculation methods

Clay bricks and clay brick masonry – Part 1, Digest DG441/1, Building Research Establishment, 1999

Clay bricks and clay brick masonry – Part 2, Digest DG441/2, Building Research Establishment, 1999

Clay bricks and clay brick masonry – Part 3, Digest DG441/3, Building Research Establishment, 1999

Contaminated land: ingress of organic vapours into buildings, Digest DG 482, Building Research Establishment, 2004

Conventions for U-value calculations, 2nd edn, B. Anderson, BR 443, Building Research Establishment, 2006

Design guide to single ply roofing, 2nd edn, Single Ply Roofing Association, 2007

Digest DG145, Heat losses through ground floors, Building Research Establishment, 1972

Digest DG180, Condensation in roofs, Building Research Establishment, 1975

Digest DG240, Low-rise buildings on shrinkable clay soils Part 1, Building Research Establishment, 1993

Digest DG241, Low-rise buildings on shrinkable clay soils Part 2, Building Research Establishment, 1993

Digest DG245, Rising damp in walls: diagnosis and treatment, Building Research Establishment, 2007

Digest DG273, Perforated clay bricks, Building Research Establishment, 1983

Digest DG293, Improving the sound insulation of separating floors and walls, Building Research Establishment, 1985

Digest DG295, Stability under wind load of loose-laid external roof insulation boards, Building Research Establishment, 1985

Digest DG298, Low-rise building foundations: the influence of trees in clay soils, Building Research Establishment, 1999

Digest DG299, Dry rot: its recognition and control, Building Research Establishment, 1993

Digest DG307, Identifying damage by wood-boring insects, Building Research Establishment, 2003

Digest DG320, Fire doors, Building Research Establishment, 1988

Digest DG327, Insecticidal treatments against wood-boring insects, Building Research Establishment, 1992

Digest DG337, Sound insulation: basic principles, Building Research Establishment, 1994

Digest DG340, Choosing wood adhesives, Building Research Establishment, 1989

Digest DG345, Wet rots: recognition and control, Building Research Establishment, 1989

Digest DG360, Testing bond strength of masonry, Building Research Establishment, 1991

Digest DG362, Building mortars, Building Research Establishment, 1991

Digest DG364, Design of timber floors to prevent decay, Building Research Establishment, 1991

Digest DG369, Interstitial condensation and fabric degradation, Building Research Establishment, 1992

Digest DG375, Wood-based panel products: their contribution to the conservation of forest resources, Building Research Establishment, 1992

Digest DG377, Selecting windows by performance, Building Research Establishment, 1992

Digest DG379, Double glazing for heat and sound insulation, Building Research Establishment, 1993

Digest DG380, Damp proof courses, Building Research Establishment, 1993

Digest DG404, PVCu windows, Building Research Establishment, 1995

Digest DG407, Timber for joinery, Building Research Establishment, 1995

Digest DG412, Desiccation in clay soils, Building Research Establishment, 1996

Digest DG416, Specifying structural timber, Building Research Establishment, 1996

Digest DG417, Hardwoods for construction and joinery: current and future sources of supply, Building Research Establishment, 1996

Digest DG423, The structural use of wood-based panels, Building Research Establishment, 1997

Digest DG429, Timbers: their natural durability and resistance to preservative treatment, Building Research Establishment, 1998

Digest DG436, Wind loading on buildings – Part 1: brief guidance for using BS 6399-2:1997, Building Research Establishment, 1999

Digest DG445, Advances in timber grading, Building Research Establishment, 2000

Digest DG461, Corrosion of metal components in walls, Building Research Establishment, 2001

Digest DG465, U-values for light steel frame construction, Building Research Establishment, 2002

Digest DG470, Life cycle impacts of timber: a review of the environmental impacts of wood products in construction, Building Research Establishment, 2002

Digest DG471, Low-rise building foundations on soft ground, Building Research Establishment, 2002

Digest DG476, Guide to machine strength grading of timber, Building Research Establishment, 2003

Digest DG492, Timber grading and scanning, Building Research Establishment, 2005

Digest DG494, Using UK-grown Douglas fir and larch timber for external cladding, Building Research Establishment, 2005

Digest DG496, Timber frame: a guide to the construction process, Building Research Establishment, 2005

Digest DG499, Designing roofs for climate change: modifications to good practice guidance, Building Research Establishment, 2006

Digest DG500, Using UK-grown Sitka spruce for external cladding, Building Research Establishment, 2006

Digest DG514, Drying distortion of timber, Building Research Establishment, 2010

Directive 2010/31/EU, the European Union, 2010

Domestic floors: repairing or replacing floors and flooring – Part 1, Construction, insulation and damp proofing, Good Building Guide GG28/1, Building Research Establishment, 1997

Domestic floors: repairing or replacing floors and flooring – Part 2, Concrete floors, screeds and finishes, Good Building Guide GG28/2, Building Research Establishment, 1997

Domestic floors: repairing or replacing floors and flooring – Part 3, Timber floors and decks, Good Building Guide GG28/3, Building Research Establishment, 1997

Domestic floors: repairing or replacing floors and flooring – Part 4, Magnesite tiles, slabs and screeds, Good Building Guide GG28/4, Building Research Establishment, 1997

Domestic floors: repairing or replacing floors and flooring – Part 5, Wood blocks and suspended timber, Good Building Guide GG28/5, Building Research Establishment, 1997

Eurocode 5 span tables: for solid timber members in floors, ceilings and roofs for dwellings, Timber Research and Development Association, 2009

External timber cladding, 2nd edn, P. Hislop, Timber Research and Development Association, 2007

Flat roof design – Part 1, Bituminous roofing membranes, Digest DG419, Building Research Establishment, 1996

Flat roof design – Part 2, The technical options, Digest DG312, Building Research Establishment, 1986

Flat roof design – Part 3, Thermal insulation, Digest DG324, Building Research Establishment, 1987

Good Building Guide GG8, Bracing trussed rafter roofs, Building Research Establishment, 1991

Good Building Guide GG16, Erecting, fixing and strapping trussed rafter roofs, Building Research Establishment, 1993

Good Building Guide GG18, Choosing external rendering, Building Research Establishment, 1994

Good Building Guide GG21, Joist hangers, Building Research Establishment, 1996

Good Building Guide GG23, Assessing external rendering for replacement or repair, Building Research Establishment, 1995

Good Building Guide GG25, Buildings and Radon, Building Research Establishment, 1996

Good Building Guide GG28 – Part 2, Concrete floors, screeds and finishes, Building Research Establishment, 1997

Good Building Guide GG33, Building damp-free cavity walls, Building Research Establishment, 1999

Good Building Guide GG36, Building a new felted flat roof, Building Research Establishment, 1999

Good Building Guide GG37, Insulating roofs at rafter level: sarking insulation, Building Research Establishment, 2000

Good Building Guide GG41, Installing wall ties, Building Research Establishment, 2000

Good Building Guide GG45, Insulating ground floors, Building Research Establishment, 2001

Good Building Guide GG47, Level external thresholds: reducing moisture penetration and thermal bridging, Building Research Establishment, 2001

Good Building Guide GG50, Insulating solid masonry walls, Building Research Establishment, 2002

Good Building Guide GG51, Ventilated and unventilated cold pitched roofs, Building Research Establishment, 2002

Good Building Guide GG60, Timber frame construction: an introduction, Building Research Establishment, 2004

Good Building Guide GG 68, Installing thermal insulation – Parts 1 & 2, Good site practice, Building Research Establishment, 2006

Good Building Guide GG74, Radon protection for new dwellings, Building Research Establishment, 2008

Good Repair Guide GR1, Cracks caused by foundation movement, Building Research Establishment, 1996

Good Repair Guide GR2, Damage to buildings caused by trees, Building Research Establishment, 1996

Good Repair Guide GR12, Wood rot: assessing and treating decay, Building Research Establishment, 1997

Information Paper IP1/00, Airtightness in UK dwellings, Building Research Establishment, 2000

Information Paper IP1/03, European standards for wood preservatives and treated wood, Building Research Establishment, 2003

Information Paper IP1/06, Assessing the effects of thermal bridging at junctions and around openings, Building Research Establishment, 2006

Information Paper IP2/05, Modelling and controlling interstitial condensation in buildings, Building Research Establishment, 2005

Information Paper IP4/01, Reducing impact and structure-borne sound in buildings, Building Research Establishment, 2001

Information Paper IP4/07, Environmental weightings: their use in the environmental assessment of construction products, Building Research Establishment, 2007

Information Paper IP4/09, Delivering sustainable development in the built environment, Building Research Establishment, 2009

Information Paper IP5/06, Modelling condensation and airflows in pitched roofs, Building Research Establishment, 2006

Information Paper IP7/95, Bituminous roofing membranes: performance in use, Building Research Establishment, 1995

Information Paper IP7/00, Reclamation and recycling of building materials: industry position report, Building Research Establishment, 2000

Information Paper IP7/01, Local authorities' performance on sustainable construction, Building Research Establishment, 2001

Information Paper IP7/08, Cements with lower environmental impact, Building Research Establishment, 2008

Information Paper IP8/08, Determining the minimum thermal resistance of cavity closers, Building Research Establishment, 2008

Information Paper IP10/10, SAP for beginners, Building Research Establishment, 2010

Information Paper IP11/00, Ties for masonry walls: a decade of development, Building Research Establishment, 2000

Information Paper IP11/10, Sustainability in foundations, Building Research Establishment, 2010

Information Paper IP13/01, Preservative treated timber: the UK's code of best practice, Building Research Establishment, 2001

Information Paper IP13/10, Cool roofs and their application in the UK, Building Research Establishment, 2010

Information Paper IP14/98, Blocks with recycled aggregates: beam and block floors, Building Research Establishment, 1998

Information Paper IP14/01, Durability of timber in ground contact, Building Research Establishment, 2001

Information Paper IP14/02, Dealing with poor sound insulation between new dwellings, Building Research Establishment, 2002

Information Paper IP16/03, Proprietary renders, Building Research Establishment, 2003

Information Paper IP18/01, Blastfurnace slag and steel slag: their use as aggregates, Building Research Establishment, 2001

Information Paper IP19/01, The performance of fibre cement slates: effects of condensation, Building Research Establishment, 2001

Insulating masonry cavity walls – Part 1, Techniques and materials, Good Building Guide GG44/1, Building Research Establishment, 2001

Insulating masonry cavity walls – Part 2, Principal risks and guidance, Good Building Guide GG44/2, Building Research Establishment, 2001

Low-rise buildings on fill – Part 1, Classification and load carrying characteristics, Digest DG 427/1, Building Research Establishment, 1997

Low-rise buildings on fill – Part 2, Site investigation, ground movement and foundation design, Digest DG 427/2, Building Research Establishment, 1998

Low-rise buildings on fill – Part 3, Engineered fill, Digest DG 427/3, Building Research Establishment, 1998

Managing Health and Safety in Construction, the Health and Safety Executive, 2007

Mastic asphalt roofing – technical guide, Mastic Asphalt Council, 2002

Metsec SFS Specification manual, Metsec, 2011

NHBC Standards, National House Building Council, 2011

PAS 23-1:1999 General performance requirements for door assemblies. Single leaf, external door assemblies to dwellings

Plasterboard – Part 1, Types and their applications, Good Building Guide GG70/1, Building Research Establishment, 2007

Plasterboard – Part 2, Fixing and finishing non-separating walls and floors, Good Building Guide GG70/2, Building Research Establishment, 2007

Plasterboard – Part 3, Fixing and finishing separating and compartment walls and floors, Good Building Guide GG70/3, Building Research Establishment, 2007

Plastering and internal rendering – Part 1, Design and specification, Good Building Guide GG65/1, Building Research Establishment, 2005

Plastering and internal rendering – Part 2, Workmanship, Good Building Guide GG65/2, Building Research Establishment, 2005

Robust details – Limiting thermal bridging and air leakage – robust construction details for dwellings and similar buildings (with amendments), Building Research Establishment, 2002

Room in roof, Product Data Sheet 1, the Trussed Rafter Association, 2007

Simple foundations for low-rise housing – Part 1, Site investigation, Good Building Guide GG39/1, Building Research Establishment, 2000

Simple foundations for low-rise housing – Part 2, 'Rule of thumb' design, Good Building Guide GG39/2, Building Research Establishment, 2001

Simple foundations for low-rise housing – Part 3, Groundworks: getting it right, Good Building Guide GG39/3, Building Research Establishment, 2002

Site investigation for low rise building – Part 1, Desk Studies, Digest DG 318, Building Research Establishment, 1987

Site investigation for low rise building – Part 2, Procurement, Digest DG 322, Building Research Establishment, 1987

Site investigation for low rise building – Part 3, Soil description, Digest DG 383, Building Research Establishment, 1993

Site investigation for low rise building – Part 4, The walk-over survey, Digest DG 348, Building Research Establishment, 1989

Site investigation for low rise building – Part 5, Trial pits, Digest DG 381, Building Research Establishment, 1993

Site investigation for low rise building – Part 6, Direct investigations, Digest DG 411, Building Research Establishment, 1995

Site-cut pitched timber roofs – Part 1, Design, Good Building Guide GG52/1, Building Research Establishment, 2002

Site-cut pitched timber roofs – Part 2, Construction, Good Building Guide GG52/2, Building Research Establishment, 2002

Span tables for solid timber members in floors, ceilings and roofs (excluding trussed rafter roofs) for dwellings, Timber Research and Development Association, 2008

Special Digest SD1, Concrete in aggressive ground, Building Research Establishment, 2005

Special Digest SD2, Timber frame dwellings. Conservation of fuel and power: AD L1A guidelines, 2nd edn, Building Research Establishment, 2006

Special Digest SD3, HAC concrete in the UK: assessment, durability management, maintenance and refurbishment, Building Research Establishment, 2002

Special Digest SD7, Insulation of timber-frame construction: U-values and regulations for the UK, Republic of Ireland and Isle of Man, Building Research Establishment, 2008

Strutting in timber floors, Wood Information Sheet 1-41, Timber Research and Development Association, 2011

The Building Regulations 2010, HMSO, 2010

The Building Regulations 2010 Approved Document A – Structure 2004 edition incorporating 2010 amendments, NBS, 2010

The Building Regulations 2010 Approved Document B (Fire Safety) – Volume 1: Dwelling houses 2006 edition incorporating 2010 amendments, NBS, 2010

The Building Regulations 2010 Approved Document C – Site preparation and resistance to contaminants and moisture 2004 edition incorporating 2010 amendments, NBS, 2010

The Building Regulations 2010 Approved Document D – Toxic substances 1992 edition incorporating 2002 and 2010 amendments, NBS, 2010

The Building Regulations 2010 Approved Document E – Resistance to the passage of sound 2003 edition incorporating 2004 and 2010 amendments, NBS, 2010

The Building Regulations 2010 Approved Document F – Ventilation 2010 edition incorporating 2010 further amendments, NBS, 2010

The Building Regulations 2010 Approved Document K – Protection from falling, collision and impact 1998 edition incorporating 2010 amendments, NBS, 2010

The Building Regulations 2010 Approved Document L1A – Conservation of fuel and power (New dwellings) 2010 edition, NBS, 2010

The Building Regulations 2010 Approved Document M – Access to and use of buildings 2004 edition incorporating 2010 amendments, NBS, 2010

The Building Regulations 2010 Approved Document N – Glazing – safety in relation to impact, opening and cleaning 1998 edition incorporating 2010 amendments, NBS, 2010

The Building Regulations 2010 Approved Document to support Regulation 7 – Material and workmanship 1999 edition incorporating 2000 amendments, NBS, 2010

The Code for Sustainable Homes, Department for Communities and Local Government, 2006

The Construction (Design and Management) Regulations 2007, HMSO, 2007

The Housing Act, HMSO, 2004

The Planning Act, HMSO, 2008

The White Book, British Gypsum, 2009

Tiling and slating pitched roofs – Part 1, Design criteria, underlays and battens, Good Building Guide 64/1, Building Research Establishment, 2005

Tiling and slating pitched roofs – Part 2, Plain and profiled clay and concrete tiles, Good Building Guide 64/2, Building Research Establishment, 2005

Tiling and slating pitched roofs – Part 3, Natural and man-made slates, Good Building Guide 64/3, Building Research Establishment, 2005

Timber frame construction, 5th edn, H. Twist & R. Lancashire, Timber Research and Development Association, 2011

Timber joist and deck floors – avoiding movement, Wood Information Sheet, Timber Research and Development Association, 1995

TRA Handbook, the Trussed Rafter Association, 2007

Trenching Practice, 2nd edn, CIRIA Report 97, CIRIA, 1992

Wood-based panels – Part 1, Oriented strand board (OSB), Digest DG477/1, Building Research Establishment, 2003

Wood-based panels – Part 2, Particleboard (chipboard), Digest DG477/2, Building Research Establishment, 2003

Wood-based panels – Part 3, Cement-bonded particleboard, Digest DG477/3, Building Research Establishment, 2003

Wood-based panels – Part 4, Plywood, Digest DG477/4, Building Research Establishment, 2003

Wood-based panels – Part 5, Medium density fibreboard (MDF), Digest DG477/5, Building Research Establishment, 2004

Wood-based panels – Part 6, Hardboard, medium board and softboard, Digest DG477/6, Building Research Establishment, 2004

Wood-based panels – Part 7, Selection, Digest DG477/7, Building Research Establishment, 2004

Wood-boring insect attack – Part 1, Identifying and assessing damage, Good Repair Guide GR13/1, Building Research Establishment, 1998

Wood-boring insect attack – Part 2, Treating damage, Good Repair Guide GR13/2, Building Research Establishment, 1998

Wood flooring, 3rd edn, P. Kaczmar, Timber Research and Development Association, 2009

Wood windows: designing for high performance, 3rd edn, P. Hislop, Timber Research and Development Association, 2009

Index